倍斯特出版事業有限公司
Best Publishing Ltd.

國貿英語
E-mail 有一套

U0066411

MP3

『抄』、『貼』加上巧思，輕鬆完成任務
E-mail 一套走天下

一套就夠用

① 掌握E-mail撰寫方式　② 仿效英文書信這樣寫
③ 運用國貿經驗談　④ 強化翻譯寫作實力

@ Email的眉眉角角：規劃「Email就是這樣用」，教你訂單滾滾來的撰寫方式，成效符合
　　客戶、廠商期待，訂單湧進獲利大增。

@ 英文書信秘訣大公開：能即學即用，用「抄」、「貼」的方式馬上就完成交辦事項。

@ 國貿經驗大分享：補足課堂與理論中的弱項、強化實務經驗，國貿實務經驗一把罩。

@ 翻譯實力速成：前輩不藏私地傳授國貿相關翻譯技巧，國貿書信寫作和翻譯實力大躍進。

　　相信許多人都曾在E-mail英文的撰寫、英文書信表達、中英翻譯上吃了不少苦頭。原因其實回歸到語句的表達跟掌握，透過相當程度的練習，大腦中就能組織出相關的英文句子。在本書《國貿英語E-mail有一套：我靠抄貼效率翻倍、獎金加倍》中我們獨家規劃了「E-mail就是這樣寫」、「英文書信這樣寫」和「翻譯小試身手」單元，讀者能藉這幾個單元演練英文文句的表達，逐步且有系統的養成自己英文書信撰寫能力，在求職上更能表現出即戰力跟競爭力。

　　此外，翻譯的句子屏除了長句和不常使用的句子，且每句均為單句，適合所有程度的學習者。基礎篇和進階篇的翻譯句子均附有MP3錄音，可以於零碎時間反覆聽，強化自己對英文語句的熟悉度，多一道「聽」的強化，讓自己印象更深刻。書籍CD中還附了便利貼光碟，收錄了「英文書信這樣寫」的信件，適當複製或超貼裡面的文句，可以省去許多撰寫英文書信所花費的時間，讓自己在時間控管上都能更掌握得宜，有更多的時間去規劃其他事項，當個上司眼中超效率的國貿人。

INSTRUCTIONS 使用說明

CONTENTS 目次

**史上超強
精華篇章**

畢業即就業,
轉職馬上錄取。

Part 1 為國貿基礎篇,規劃
了 38 個必會國貿單元,實
務經驗能立馬提升。打好
基礎,求職、應試、轉職
不再自亂陣腳。

1 Part · 國貿基礎篇

Unit 01 · 寄送產品目錄 Delivery of Catalogue ⋯⋯⋯⋯ 016

Unit 02 · 寄發產品目錄後之追蹤(不符合買方需求)
Follow-Up after Catalogue Delivery (Not
Conforming to Buyer's Demand) ⋯⋯⋯⋯ 020

Unit 03 · 回覆還價 Feedback to Counter Offer ⋯⋯⋯⋯ 024

Unit 04 · 引薦客戶 Request of Customer
Recommendation ⋯⋯⋯⋯ 028

Unit 05 · 邀請觀展 Invitation to the Trade Exhibition ⋯⋯⋯⋯ 032

Unit 06 · 會展招呼訪客 Serving the Visitor ⋯⋯⋯⋯ 036

Unit 07 · 會展議價 Price Negotiation in the Exhibition ⋯⋯⋯⋯ 040

Unit 08 · 感謝招待 Expressing Appreciation for
Hospitality ⋯⋯⋯⋯ 044

優質經驗談

▶ 出口價格估算：

FOB 售價 = FOB 成本價+FOB 售價 x（銀行手續費率＋推廣費率＋押匯貼現率＋商港建設費率＋利潤率）

移項後得：

$$FOB\ 售價 = \frac{FOB\ 成本價}{1-（銀行手續費率＋推廣費率＋押匯貼現率＋商港建設費率＋利潤率）}$$

CFR 售價＝CFR 成本價＋CFR 售價（手續費率＋貼現率＋利潤率）＋商港建設費＋推廣費

▶ 移項後得：

$$CFR\ 售價 = \frac{CFR\ 成本價＋商港建設費＋推廣費}{1-（手續費率＋貼現率＋利潤率）}$$

CIF＝CFR 成本價＋CIF 售價×1.1×保險費率＋（手續費率＋貼現率＋利潤率）×CIF 售價＋商港建設費＋推廣費

▶ 移項後得：

CIF 售價＝ CFR
1-（1.1×保險

超實務規劃，跟**學用落差**
說再見，快點偷偷學起來吧！

Unit 16．放寬付款條件 Request of Easing Payment Term

英文書信邊寫

Notification of L/C Acceptance 信用狀承兌通知

Dear Tom,

We're writing to inform you that your P/O No. 1234 has been delivered on board on May 20 as schedule per the attached copies of the shipping documents.

In accordance with the irrevocable L/C No. L8 123 for USD 3,300 issued by London Bank, we have valued on you at sight against this shipment. The full set of shipping documents will be sent to you, once you accept the draft. Please kindly honor the draft immediately upon presentation.

Sincerely Yours,
Tony Yang

中文翻譯

湯姆 您好：

僅此文通知貴公司，號碼為 1234 的訂單已如期於五月二十日裝船，茲附上出貨文件副本。依據倫敦銀行開出之總額 3,300 美元的不可撤銷信用狀號碼 L8 123，我司公司已對此批

貨開出見票即付之匯票，一經貴司承兌後，將寄出全套裝船文件予貴公司。懇請貴公司於見票後立即予以承兌。

湯尼 楊 敬啟

077

獨家規劃 E-mail 抄貼光碟，靠抄貼就高效率完成工作，副作用就是事情永遠做比別人快 XD! 連老闆都覺得你神神 DER，當然特別喜歡**產值高**的你。

CONTENTS 目次

1 Part · 國貿基礎篇

Unit 01 · 寄送產品目錄 Delivery of Catalogue ·········· 016

Unit 02 · 寄發產品目錄後之追蹤（不符合買方需求）

Follow-Up after Catalogue Delivery (Not

Conforming to Buyer's Demand) ·········· 020

Unit 03 · 回覆還價 Feedback to Counter Offer ·········· 024

Unit 04 · 引薦客戶 Request of Customer

Recommendation ·········· 028

Unit 05 · 邀請觀展 Invitation to the Trade Exhibition ·· 032

Unit 06 · 會展招呼訪客 Serving the Visitor ·········· 036

Unit 07 · 會展議價 Price Negotiation in the Exhibition 040

Unit 08 · 感謝招待 Expressing Appreciation for

Hospitality ·········· 044

CONTENTS 目次

Unit 09 · 詢價及報價 Inquiry & Quotation 048

Unit 10 · 追蹤報價 Follow up the Quotation 052

Unit 11 · 議價不成 Failure of Counter Offer 056

Unit 12 · 調價通知 Notification of Price Adjustment 060

Unit 13 · 催促開立信用狀 Urge to Open L/C 064

Unit 14 · 要求修改信用狀 Request of L/C Amendment 068

Unit 15 · 通知銀行改信用狀 Notify Bank of L/C

Amendment 072

Unit 16 · 放寬付款條件 Request of Easing Payment

Term 076

Unit 17 · 託收付款 Payment term as Collection 080

Unit 18 · 拒絕承兌 Refuse of Acceptance 084

Unit 19 · 承兌延遲 Delay of Acceptance 088

Unit 20 · 支票付款 Payment Terms as Check ⋯⋯⋯⋯⋯ 092

Unit 21 · 退票通知 Notification for Bounced Check ⋯⋯ 096

Unit 22 · 洽詢保險公司 Enquiry to Insurance

Company ⋯⋯⋯⋯⋯⋯⋯⋯⋯⋯⋯⋯⋯⋯⋯⋯⋯⋯⋯⋯ 100

Unit 23 · 需求額外保險 Request of Additional

Insurance ⋯⋯⋯⋯⋯⋯⋯⋯⋯⋯⋯⋯⋯⋯⋯⋯⋯⋯⋯ 104

Unit 24 · 提出索賠通知 Notification of Claim ⋯⋯⋯⋯⋯ 108

Unit 25 · 洽詢船班 Enquiry of Shipping Schedule ⋯⋯⋯ 112

Unit 26 · 裝船通知 Shipping Advice ⋯⋯⋯⋯⋯⋯⋯⋯⋯ 116

Unit 27 · 漏裝處理 Make-Up for Short Delivery ⋯⋯⋯⋯ 120

Unit 28 · 數量不符 Error Shipping Quantity ⋯⋯⋯⋯⋯⋯ 124

Unit 29 · 功能瑕疵 Function Defect ⋯⋯⋯⋯⋯⋯⋯⋯⋯ 128

CONTENTS 目次

Unit 09 · 產品開發 Product Development ⋯⋯⋯⋯⋯⋯ 218

Unit 10 · 送樣審核 PPAP Verification ⋯⋯⋯⋯⋯⋯⋯ 224

Unit 11 · 邀請觀展 Invitation to the Trade Exhibition ⋯⋯ 230

Unit 12 · 會展接待 Serving the Visitors ⋯⋯⋯⋯⋯⋯ 236

Unit 13 · 展後拜訪 Visitation Plan to Potential Customer
Meeting in the Exhibition ⋯⋯⋯⋯⋯⋯⋯ 242

Unit 14 · 簽證申請 Applying for Visa ⋯⋯⋯⋯⋯⋯⋯ 248

Unit 15 · 交通安排 Transportation Arrangement ⋯⋯⋯ 254

Unit 16 · 食宿預定 Restaurant & Hotel Reservation ⋯⋯ 260

Unit 17 · 人員介紹 Introduction of Company Staff ⋯⋯ 266

Unit 18 · 公司介紹 Tour of Office Building ⋯⋯⋯⋯ 272

Unit 19 · 參觀工廠 Tour of Facility ⋯⋯⋯⋯⋯⋯⋯ 278

Unit 20 · 認證稽核 Certification ⋯⋯⋯⋯⋯⋯⋯⋯⋯ 284

Unit 21 · 活動邀請 Invitation ⋯⋯⋯⋯⋯⋯⋯⋯⋯⋯ 290

Unit 22 · 人事異動 Reshuffle Announcement ⋯⋯⋯ 296

Unit 23 · 喬遷啟示 Removal Notification ⋯⋯⋯⋯⋯ 302

Unit 24 · 表達祝賀 Expressing Congratulation ⋯⋯ 308

Unit 25 · 表達弔唁 Expressing Condolence ⋯⋯⋯⋯ 314

1 Part 國貿基礎篇

國貿基礎篇介紹了國貿人入門必備的寄送產品目錄、詢價、報價、議價、信用狀使用、付款、洽詢船班、索賠和要求換貨。單元中的 E-mail 這樣寫可強化 E-mail 撰寫跟表達。英文書信這樣寫則介紹簡易的短篇書信表達。每篇書信後加入國貿經驗談，可以大幅地增進國貿知識。最後則是翻譯小試身手，可以於單元結束後練習相關的國貿翻譯句子，強化中翻英能力。

E-mail 這樣寫

 MP3 01

❶ I'd like to inquire about your PA0925 Extendable Selfie Stick which I've seen its information on your website. I'm highly interested in learning more about it. Please let me know its price and availability status.

我想要詢問貴公司的 PA0925 伸縮自拍神器，我在您們的網站上有看到了這項產品的資訊，我很有興趣，想要多瞭解一些，還請您告訴我它的價格與現貨狀況。

❷ We're interested in your CPY1202 product and would like to know its price, availability status and also the estimated freight cost to Taiwan. Please inform. If there are any additional fees, e.g. handling fee, please tell us as well. Thanks.

我們對您的 CPY1202 產品有興趣，想知道它的價格與現貨狀況，以及貨到臺灣的預估運費，還請告知。若還有像是手續費這類得額外收取的費用，也請一併告知，謝謝。

英文書信這樣寫

Request of Catalogue 索取產品目錄

Dear Sir and Madam,

We learnt from the advertisement about your high-quality valve, which will have a potential market in our local area. We are very interested in those valves and would like to request detailed information about your product. It would be appreciated if you could provide us the latest catalogue including the price list.

Looking forward to your prompt response.

Sincerely yours,
Tom Smith

 中文翻譯

敬啟者：

我們公司從廣告中得知貴公司高品質閥門產品，此產品在我們當地將有市場潛力。我們公司對貴公司的產品非常感興趣，因此寫信需求詳細的產品訊息。如貴公司能提供最新的產品目錄同時附上報價單，我們公司不勝感激。

期待您早日回覆。

湯姆 史密斯 敬啟

Unit 02

寄發產品目錄後之追蹤（不符合買方需求）Follow-Up after Catalogue Delivery (Not Conforming to Buyer's Demand)

 E-mail 這樣寫 MP3 03

❶ Our list price for S-1000 Calcium Tablets is US$ 100 and you could receive our 40% distributor discount for a price of US$ 60 per box. Attached please find our official quote for 500 boxes. Please note that the quote is valid through December 31, 2016.

我們 S-1000 鈣片的定價為 US$ 100，您可享有 40％的代理商折扣，每盒的折扣後價格為 US$ 60。在此附上我們 500 盒的正式報價單，請注意此報價至 2016 年 12 月 31 日前有效。

❷ The quote attached reflects the transfer price which your typical distributor discount is applied. Please be sure to reference the quote at the time of order placement in order to guarantee the correct pricing.

附上的報價單所列的是移轉價格，有加計了給您們的一般代理商折扣。下單時請務必加註報價單單號，以確保我們能給您正確的價格。

英文書信這樣寫

Follow-up after Catalogue Delivery 寄發型錄後之追蹤

Dear Tom,

I am writing to ask if you received our catalogue that we sent to you last week per your request. One week has passed, but no response has been received. Just to be sure, we are enclosing a copy of our catalogue for your reference. We would be grateful if you could confirm the receipt of it by return.

We look forward to hearing from you soon.

Sincerely yours,
Tony Yang

 中文翻譯

湯姆 您好：

茲發本文詢問貴公司是否收到上週我們公司依貴公司要求所寄出之產品目錄。已過了一週，我們公司仍未收到任何回覆。為了保險起見，現附寄目錄一份供您參考。如果貴公司能回覆確認收到，我們公司將不甚感激。

期待您的儘速回覆。

湯尼 楊 敬啟

Part 1 國貿基礎篇

Part 2 國貿進階篇

03 回覆還價 Feedback to Counter Offer

 E-mail 這樣寫   MP3 05

❶ We recently instituted a volume discount program as attached to make it easier for you to get even greater discount values on your purchases.

我們最近訂定了數量折扣方案如附，讓您採購時更容易享受到更大的折扣。

❷ We do provide quantity discounts on our products. We encourage you to make your purchases in these larger quantities to enjoy the discount.

我們對產品有提供數量折扣，建議您在數量上多採購一些，以享有折扣的優惠。

英文書信這樣寫

Feedback to Counter Offer 回覆還價
Dear Tom,

Appreciate your counter offer by email, but we are sorry that we can't grant the discount due to the small size of your order. We'd be willing to offer 2% special discount based on the MOQ 5000pcs. Taking the quality into consideration, we firmly believe our price is favorable. We hope you will reconsider and accept our offer.

Sincerely yours,
Tony Yang

中文翻譯

湯姆 您好：

感謝貴公司來信還價，然而很抱歉基於貴公司的小額訂單量，我司無法提供折扣。但在能符合最小訂購量 5000 件的基礎下，我們公司可以提供 2% 的特別折扣。考量到產品品質，深信我們的價格是有利的。期望您再次考慮，並接受我們的報價。

湯尼 揚 敬啟

04

引薦客戶 Request of Customer Recommendation

 E-mail 這樣寫

❶ By ordering 2 or more vials of the same product and pack size, you will benefit from our new discount program. You can save up to 25%.

若對同產品、同包裝訂購的量達兩瓶或兩瓶以上時,就可適用我們新的折扣方案,可以替您節省高達 25%的成本。

❷ The price for KT-0710 is $1185.00 per kit FOB Seattle. Freight and handling costs are additional and depend on the quantity ordered.

每組 KT-0710 的西雅圖交貨 FOB 價格為$1185.00,運費與處理費皆須另計,金額視訂購數量而定。

英文書信這樣寫

Request of Customer Recommendation 引薦客戶推薦
Dear Tom,

Thank you for the continued support to us. As you know, our company is expanding our business oversea. We would appreciate it if you could kindly recommend some potential customers to us.

Thank you in advance for your help and assistance.

Sincerely yours,
Tony Yang

 中文翻譯

湯姆 您好:

感謝您對我們公司的持續支持。如您所知,我們公司正在擴展海外業務。如您能引薦一些淺在客戶,我們公司將不勝感激。

預先感謝您的幫忙與協助。

湯尼 楊 敬啟

國貿經驗談

▶▶ 買方在檢閱產品目錄後，可要求賣方提供部分樣品，此為檢視樣品及瞭解產品性能最直接之方式，但考量單一產品單價、產品品項多寡等因素，賣方可僅提供部分品項，或另收取樣品費用。索取樣品時可依據產品目錄陳述，告知所需之型號、顏色、材質和規格等，或直接需求賣方提供其最暢銷之款式。

▶▶ 索取樣品以小批量為主，通常為一至數十件不等，因此多會選擇快遞運輸方式。寄件時，即使外包裝已特別註明不可重摔、小心輕放等字樣，但難保快遞員會妥善處理，所以堅固內外包裝極為重要，例如使用內盒、泡棉、氣泡袋、或隔板，以避免運輸過程中相互碰撞。但切勿重重包裹外盒，例如多層膠帶包覆將造成拆解困難，想必會造成收件者的困擾。

翻譯小試身手

❶ 如果你的客戶要求 10 組都來自於同一個批次，那麼出貨準備期就會是接單後的 2～3 個星期。

❷ 請告訴我客戶需要的量有多少，這樣我就可以給你個正式的報價。

參考答案

MP3 08

❶ If your customer wants 10 kits from one single lot, the lead time would be 2-3 weeks from receipt of order.

❷ Please let me know what quantity the customer will need so that I can develop an official quote for you.

Unit
05 邀請觀展 Invitation to the Trade Exhibition

 E-mail 這樣寫 MP3 09

❶ Unfortunately, your offer is high while our customer's budget is limited. For your information, the customer has a lot of potential of placing bulk orders in the future, and also has the power to influence quite a few users in the market.

但您報來的價格高，而我們客戶的預算卻很有限。也讓您知道一下，這個客戶未來很有潛力下大筆的訂單，也有能力影響市場上的其他使用者。

❷ To compete with our arch-rival and let this key customer switch to buy from us, we hope you could support this project by offering your best bottom line price. We'll also lower our margin to get the most competitive price to the customer.

為了跟我們的主要競爭對手競爭，並為了讓這個重要客戶轉而成為我們的主顧，我們希望您可以提供支援，報給我們您的最底價，而我們也會降低我們的利潤，讓報給客戶的價格能夠最具競爭力。

英文書信這樣寫

Inquiry 詢價

Dear Tony,

We hereby acknowledge the receipt of the sample you sent to us last week, and feel quite satisfied with your product quality. We'd like to inquire about the piece price of your P/N 0007 and 0008.

We should be obliged by your early reply.

Sincerely yours,
Tom Smith

中文翻譯

湯尼 您好：

　　特此通知已查收貴公司上週來樣，我們公司相當滿意貴公司的產品品質，想詢問產品件號 *0007* 及 *0008* 之單價及付款條件。

　　如貴公司能早日回覆，我們公司將不勝感激。

湯姆 史密斯 敬啟

▶▶ 參展報名流程：線上登錄→繳納訂金→郵寄報名→資格審核→攤位規劃→確認參展資格（不合格→退回訂金／合格→開立訂金收據）→展前說明會（決定攤位位置）→繳交費用→會展相關作業申請→進場裝潢→開展

▶▶ 每年世界各地各行各業皆會舉辦大大小小展覽，選擇符合自身企業屬性的展覽要先考量該展覽是否涵蓋了您的銷售市場，並非大城市舉辦的大型展覽就是適合的地點，展覽地須能吸引潛在客戶群，而且必須是符合產品導向，例如能符合產品生產排程、展期可配合新產品上市時間，且能符合相應的產品廣告等，畢竟參展費用可是一筆可觀的數目。

翻譯小試身手

❶ 若是您只買 3 組，就不會有特價，而是適用您平常所享的代理商折扣。

_____

_____

❷ 請您指明會用哪一種貨幣（美金、歐元或英鎊）來支付這筆交易的貨款。

_____

_____

參考答案

 MP3 10

❶ If you purchase only 3 kits, there will be no special price and your usual distributor discount would apply.

❷ Please specify the currency (USD, EUR or GBP) which you will be using to pay for this transaction.

Unit
06
會展招呼訪客
Serving the Visitor

 E-mail 這樣寫

❶ As a result of an increase in the pricing of our raw materials and production costs of our products, we are here to inform you of our decision to increase prices by 3% from January 1st 2017.

我們產品的原料和生產成本都提高了，因此在此通知您，我們已決定將我們的價格調高 3%，自 2017 年 1 月 1 日起生效。

❷ We have been in production for over 10 years but not had a widespread price adjustment. Effective January 15, 2017, we'll implement a price increase of 5% across our whole range of products.

我們投入生產已超過 10 年了，但價格都還沒有大規模調整過。從 2017 年 1 月 15 日開始，我們所有產品的價格都將調漲 5%。

英文書信這樣寫

Follow up the Quotation 追蹤報價

Dear Tom,

Through the email, we wish to follow up the quotation sent to you via email dated Feb. 1, 2014. I've attached the quotation sheet again, just in case you didn't receive it. We are very eager to know if you'd like to proceed with the order if the terms are acceptable. Please place the order at your earliest convenience and we'll arrange to carry out the production immediately.

We truly believe our valve to be of high quality and with high functionality and are sure the production will meet your requirements.

We are anxious to receive your order.

Sincerely yours,
Tony Yang

湯姆 您好：

　　藉由此信追蹤我司在 2014 年 2 月 1 日寄給您的報價。我再次附上報價單，以防您未收到。我們公司很想知道貴公司是否想要啟動生產此筆訂單。如果這些條件都可接受，請儘早下訂單，我們將立即排產。

　　深信我們公司的閥門品質高檔且高功能，保證能符合您的需求。

　　我們渴望收到貴公司的訂單。

湯尼 揚 敬啟

國貿經驗談

▶ 參加國外展覽是行銷計畫的一環，由於展覽的準備期很長，在決定參加某一展覽之後，最好成立工作小組或指定專人負責規劃。工作小組除負責與主辦單位連絡外，亦同時負責推動部分參展工作、控制進度及各單位間之協調配合。展前準備工作必須配合主辦單位之進度，依照展覽舉辦頻率的不同，最早可能在展出前一年多就必須開始，比較晚者，至少亦須於展前半年推動。

▶ 詳細內如請參閱「廠商赴國外參展標準流程（SOP）及應注意事項：中華民國對外貿易發展協會 98.11.25 編撰」。

▶ 展期通常不超過一週，如何在短時間達到最高經濟效益，是參展的最主要目的，因此參展人員扮演舉足輕重的腳色，進行展前的

培訓及準備工作相當重要，尤其針對首次參與展覽的新生菜鳥。培訓須著重在對產品的熟悉度，如產品規格、型號、顏色等。如能現場實際操作的產品，也須清楚產品性能及如何操作，並要瞭解各展品在攤位中的相關擺設位置。

▶▶ 參展人員需在展館中與陌生人交談、介紹產品、發送印刷品或精品，因此身為參展人的您若能具備樂於跟陌生人交談的個性，必能更駕輕就熟。

翻譯小試身手

❶ 我很樂意在您快要下單前，提供現有批次到期日的資訊給您。

_____

_____

❷ 此報價一個月內有效。

_____

_____

參考答案

MP3 12

❶ I will be happy to give you the expiration information of available lots when you are closer to placing the order.

❷ This quotation will expire in one month.

Unit

07 會展議價 Price Negotiation in the Exhibition

 E-mail 這樣寫

MP3 13

❶ Please note that this product is a limited quantity dangerous item. To ship this product internationally, there will be an $80 handling fee.

要請您注意一點，這項產品屬於限制數量的危險品項，要辦理出口則須收取$80 的處理手續費。

❷ If you choose to ship via FedEx freight collect, you'll be responsible for the freight charges.

若您選擇 FedEx 運費到付的方式來出貨，那麼運費就是由您來負擔。

英文書信這樣寫

Counter Offer 還價

Dear Tony,

We appreciate your quotation offered to us on Feb. 1, 2014. Unfortunately, your price is too high for us to consider and will leave us no margin of profit on our sales if we were to accept the price you quote.

In consideration of fact that we have been doing business with each other for many years and our order is increasing every year. Please consider offering 3% discount, and that will be an acceptable deal. We sincerely hope you can accept our request and confirm by return soon.

Sincerely yours,
Tom Smith

湯尼 您好：

　　感謝您 2014 年 2 月 1 日提供的報價。很可惜，貴公司的價格太高，超出我們所能考慮的範圍，如果以此價格購買產品，我們的銷售將無利潤。

　　有鑒於雙方生意往來多年，且我們公司的訂單逐年成長。請考慮給予我們公司 3%的折扣，如此這將是一筆可行的交易。我們公司真誠的希望貴司能接受我們的要求，並儘速回覆確認。

湯姆 史密斯 敬啟

國貿經驗談

電匯（Telegraphic Transfer, T/T）

▶▶ 電匯是目前國際貿易中使用最廣的一種匯款方式，電匯是指匯出行應匯款人的申請，以電訊方式將電匯付款委托書給匯入行，指示解付一定金額給收款人的一種匯款方式。電匯方式的優點是收款人可迅速收到匯款，但費用較高。常見的電訊方式有 SWIFT，電傳（TELEX），電報（CABLE, TELEGRAM）等。

▶▶ 在國際貿易實務中可分為前 T/T（預付貨款）和後 T/T（裝船後或收貨後付款）兩種。比起信用狀交易，電匯的風險仍高一些，但就銀行手續費來說，電匯費用要比信用狀費用低許多。

▶▶ 電匯的流程：

- 匯款人填寫電匯申請書並交款付費給匯出行
- 由匯出行拍發加押電報或電傳給匯入行
- 匯入行發電匯通知書給收款人
- 收款人接到通知後至銀行兌付
- 銀行解付款項給收款人
- 解付完畢，匯入行發出借記通知書給匯出行，同時匯出行發電匯回執予匯款人

翻譯小試身手

❶ 這項產品每 1kg 的代理商單價為 1,500 歐元，另外，對於國際匯款，我們還會加收 30 歐元的國際匯款轉帳的銀行手續費。

❷ 一次下單 10 組或更多量時，才能適用此價格。

參考答案 MP3 14

❶ The product's distributor price is EUR 1.500,00 per 1 kg. There will also be an extra EUR 30,00 fee to cover bank charges incurred from international wire transfer.

❷ This pricing is only valid on the single order of 10 kits or more.

 E-mail 這樣寫 MP3 15

❶ We have checked on the bank charges we have paid so far and it has varied between USD 18.00 to 35.00.

我們查了先前付過的銀行手續費，發現金額從 USD 18.00 到 35.00 都有。

❷ We ask that a distributor take a very small margin (3%) to process this kind of projects as we are also cutting our margin to accommodate key customers.

對於這類案子，我們要求代理商將利潤率降到很低（3%）來做，因為我們也同時降低了利潤率來配合重要客戶的要求。

英文書信這樣寫

Acceptance of Quotation 價格確認

Dear Tony,

We're pleased to accept your offer by email dated Feb. 5, 2014. Please see the attached order sheet and acknowledge receipt by signing back with delivery date. As we need the parts urgently, it would be grateful if you could arrange the shipment by this weekend.

Since the matter is urgent, an early reply will oblige.

Sincerely yours,
Tom Smith

 中文翻譯

湯尼 您好：

　　我們公司很高興接受貴司 2014 年 2 月 5 日的報價。請查閱附件訂單，並註記出貨日後回簽，以確認貴公司收到此訂單。因我們公司緊急需要此批產品，如可於本週末前出貨，我們公司將不勝感激。

　　既然此事緊急，請儘早給予答覆。

湯姆 史密斯 敬啟

Unit

09

詢價及報價
Inquiry & Quotation

 E-mail 這樣寫   MP3 17

❶ The transfer price to you this time will be $1,200 per set while the end user price needs to be $1,290 per set. Please let us know if you accept this request.

我們這次給您的移轉價格會是每組$1,200，而您給客戶的價格須為每組$1,290，請告知您是否接受這樣的要求。

❷ Attached please find the distributor price list in EUR. On the mentioned list prices you will get a 50% discount (these are all transfer prices without any further discount).

在此附上歐元的代理商價格表，就所列出的定價，您們可享 50％的折扣（這些皆為移轉價格，不再有任何額外的折扣）。

英文書信這樣寫

Submit Draft Contract 提出草約

Dear Tony,

According to our discussion, enclosed is our draft contract in duplicate for your reference.

Please check the terms of the draft and keep us informed if you have any concern. We can have a call conference to further discuss the subject if needed.

We await your reply soon.

Sincerely yours,
Tom Smith

中文翻譯

湯尼 您好：

　　依據我們討論內容，請參閱附件草約副本。

　　請檢視草約中的條款，並告知貴公司對草約內容是否有其它顧慮。如果有需要，我們可以進行電話會議進一步討論此議題。.

　　等待您的儘速回覆。

湯姆 史密斯 敬啟

▶▶ 貨櫃（container）一般有固定的收費標準，以確保船公司在運輸過程中支出全部費用後，均能得到相應的補償。船公司通常須支付貨櫃貨物在運輸過程中所產生的全部費用，包括海上運輸費用、內陸運輸費用、各種裝卸費用、搬運費、手續費、服務費等。

▶▶ 國際貿易出貨方式分，簡單來說可分為兩大類，整櫃貨或併櫃貨，整櫃貨指「同一」出口商產品裝載於同一貨櫃並運送至同一交貨地；而併櫃貨指「不同」出口商產品裝載於同一貨櫃運送至同一目的港後，再發貨至不同地點之各別客戶。而併櫃貨運費計算方式則分為重量噸計價貨體積噸計價，計算公式如下：

▶▶ 貨物體積重之換算：

重量噸：1 Ton（噸）＝1,000kgs

體積噸：1 M3＝1CBM＝1 立方米

▶▶ 計算方法：

1 M＝100 cm，1 M3＝100 cm×100 cm×100 cm

1 M3＝1 Cubic Meter＝1 CBM，1 材＝1 Cubic Feet（英制）＝1 Feet3 簡寫為 1'

＝1 feet×1fee×1feet（英呎）3＝12"×12"×12"＝1728 吋 3

＝30.48 cm×30.48 cm×30.48 cm＝（0.3048M）3

＝0.02831684659M3

∴1M3＝35.3146667239 才（35.315）

（簡單的計算方式）

每一外箱的長（CM）×寬（CM）×高（CM）÷28317（公式）

＝材

材×箱數÷35.315＝CBM (M3)

翻譯小試身手

❶ 附上我們對最小標準包裝產品的一般報價。

_____

_____

❷ 這項產品可從我們的固定存貨來供貨，價格為$250.00。

_____

_____

參考答案

MP3 18

❶ Attached is a general quote for our smallest standard pack size.

❷ This product can be purchased from our regular inventory for $250.00.

 E-mail 這樣寫
 MP3 19

❶ Thank you for your inquiry for bulk sizes of Item # 1202. Attached is our quotation to your company. Please note that this quotation will expire on 04/23/17. If I can help with anything else, please just let me know.

謝謝您發來產品# 1202 大包裝的詢價，在此附上我們給您公司的報價單，請注意此報價單的效期至 04/23/17 為止，若還有什麼需要我協助的地方，就請直接告訴我。

❷ I have attached the quote. Please make sure to reference the quote no. on your PO in order to receive the right discount.

我已附上了報價單，下單時請您務必在訂單上標註報價單單號，以能拿到正確的折扣。

英文書信這樣寫

Place Purchase Order 下採購單

Dear Tony, we are satisfied with your offer by email dated March, 23, 2014. Hereby attachment is our PO number 1234. Please be informed that we are in urgent need of the goods. Your prompt attention to this order with earliest delivery will be greatly appreciated.

We hope to receive a prompt reply per return mail.

Sincerely yours,
Tom Smith

中文翻譯

湯尼 您好：

　　我們公司非常滿意貴公司於 2014 年 3 月 23 日信函中所提供的報價。在此附上號碼為 1234 的訂單。 在此通知我們公司急需此批貨。如果貴公司能儘速處理此訂單並安排最快交貨，我們公司將不甚感激。

　　我們希望收到貴公司盡速回函。

湯姆 史密斯 敬啟

Unit

11 議價不成 Failure of Counter Offer

 E-mail 這樣寫 MP3 21

❶ We'd like to place an order for your Controller. Attached please find our Purchase Order. Please confirm and process the order for us. Also please e-mail your Proforma Invoice to me so that we could make payment to you by wire transfer.

我們想要下單訂購您的控制器,在此附上訂購單,還請確認並替我們處理此訂單,也請 e-mail 形式發票給我,這樣我們才能安排用電匯來支付貨款給您。

❷ Following our previous discussions, we are pleased to place our initial order to you for 10 kits of PY1202 Controller. The product details and our preferred shipping method are listed as follows. Please confirm the order and let us know the estimated shipping date.

延續我們先前的討論,我們很高興要來跟您下我們的第一張訂單,訂購 10 組的 PY1202 控制器。在此列出產品明細以及我們想要的出貨方式,請您確認此訂單,並告知預估的出貨日。

英文書信這樣寫

Part 1 國貿基礎篇

Part 2 國貿進階篇

Order Change 更改訂單

Dear Tony,

May we ask your attention to our requirement of quantity increase to 1000 pieces per each item via PO number 1234? The quantity change is due to the market demand, higher than we expected

We'll send the revised order by e-mail. Please confirm to see whether the delivery date remains unchanged. Besides, due to the larger order size, I'd inquire to see if there is any room for a downward regulation of piece price.

Please favor us with your reply as early as possible.

Yours sincerely,
Tom Smith

湯尼 您好：

　　可否請求貴公司注意我們需求號碼為 1234 的訂單其每個品項增加數量到 1000 件？訂單數量的改變源自於市場需求增加高於我們預期。

　　我們會電郵寄出修改後的訂單。請回覆確認交期是否維持不變。此外，由於訂單數量較大，想詢問看看是否產品單價有調降的空間。

　　請協助儘速回覆。

湯姆 史密斯 敬啟

國貿經驗談

▶▶ **進口商售價的計算：**

進口商售價＝總成本＋售價×（營業加值稅率＋預期利潤率）
總成本＝CIF＋進口費用

移項後：
售價[1－（營業加值稅率＋預期利潤率）]＝總成本

再移項：
售價＝總成本／1－（營業加值稅率＋預期利潤率）

成本是指產品在生產和銷售的過程中，所需支出的相關費用，如生產設備折舊、中間產品消耗、勞動支出等其它開銷的加總。而在國際貿易實務上，進口產品的成本就須包含生產和銷售該產品的實際成本，加上相關利潤。低於正當成本的進口商品價格，稱為傾銷價格；相反的，高於進口商品價格則是壟斷價格。

翻譯小試身手

❶ 我們需要一份證明，說明這項產品將僅供目的地國家境內研究用。

_____

_____

❷ 這項產品有現貨，單價為美金$595，不包含運費。

_____

_____

參考答案 MP3 22

❶ We need a certification stating that the product is to be applied exclusively for research use within the country of destination.

❷ This product is available and the price is $595 USD per kit excluding shipping charges.

 E-mail 這樣寫 MP3 23

❶ Please register on our website, login and visit the 'My Account' page to complete credit account form and email it to us along with a blank copy of your company headed paper. Once your account has been set up, we will advise you and you will be able to place online orders.

請先在網站上登記，登入後請到「我的帳戶」填寫信用帳戶格式，再連同有您公司抬頭的空白信紙一起 e-mail 給我們。等您的帳戶建立了之後，我們會通知您，您就可以線上下單了。

❷ Please consider ordering online. It allows you to track the status and shipment of your order, re-order from your order history and retrieve application news relevant to the ordered products.

請考慮線上下單的方式，它可讓您追蹤訂單的狀況與出貨情形，也可從歷史訂單中直接重新訂單，還能直接取得所訂產品的相關應用資訊。

英文書信這樣寫

Order Rejection 拒絕訂單

Dear Tom,

We are in receipt of your order number 3344 sent by email dated March 30. To our greatest regret we must herewith inform you that we cannot at present entertain any new orders, due to the tight production capacity. We are, however, keeping your order before us. As soon as we are in a position to accept new orders, we will send you an email immediately. Let me reiterate our sincere regret regarding these problems. Your kind understanding of our situation will be appreciated.

Sincerely yours,
Tony Yang

中文翻譯

湯姆 您好：

　　我司收到貴公司 3 月 30 日電郵寄出號碼為 3344 的訂單。非常遺憾通知貴公司，由於產能吃緊，我們公司需在此通知無法接受任何新訂單。然而，我們會保留貴司訂單，如果可以開始接單時，我們公司將立即電郵通知貴公司。請容找再次因此問題至上誠摯的歉意。貴公司能理解我們公司的立場，將不甚感激。

湯尼 楊 敬啟

國貿經驗談

▶▶ 出口價格估算：

FOB 售價 = FOB 成本價+FOB 售價 x（銀行手續費率＋推廣費率＋押匯貼現率＋商港建設費率＋利潤率）

移項後得：

FOB 售價＝ $\dfrac{\text{FOB 成本價}}{1-（\text{銀行手續費率＋推廣費率＋押匯貼現率＋商港建設費率＋利潤率}）}$

CFR 售價＝CFR 成本價＋CFR 售價（手續費率＋貼現率＋利潤率）＋商港建設費＋推廣費

▶▶ 移項後得：

CFR 售價＝ $\dfrac{\text{CFR 成本價＋商港建設費＋推廣費}}{1-（\text{手續費率＋貼現率＋利潤率}）}$

CIF 售價＝CFR 成本價＋CIF 售價×1.1×保險費率＋（手續費率＋貼現率＋利潤率）×CIF 售價＋商港建設費＋推廣費

▶▶ 移項後得：

CIF 售價 ＝ $\dfrac{\text{CFR 成本價＋商港建設費＋推廣費}}{1-（1.1×\text{保險費率＋手續費率＋貼現率＋利潤率}）}$

Part **1** 國貿基礎篇

Part **2** 國貿進階篇

翻譯小試身手

❶ 請告訴我們出貨包裝的毛重與尺寸，我們要跟這裡快遞公司報的運費比較一下。

❷ 貨運的落地成本總額包含了產品買價、運費、保險，以及抵達目的地機場／港口前的所有其他成本。

參考答案 MP3 24

❶ Please give us the package's gross weight and dimension for us to check the freight with our courier agent.

❷ The total cost of a landed shipment includes purchase price, freight, insurance, and other costs up to the port of destination.

Unit

13 催促開立信用狀 Urge to Open L/C

 E-mail 這樣寫

❶ Add items to your shopping cart by clicking the "buy" button next to your desired product. You are then directed to the shopping cart page.

點選您所要產品旁的「購買」按鍵，加入購物車，則會引導您至購物車的頁面。

❷ You can select to proceed to checkout, alternatively you can choose to continue shopping. If you select to go to checkout, you will be directed to the shipping info page.

您可選擇進行結帳，也可選擇繼續購物。若您選擇結帳，則會引導您至出貨資訊的頁面。

 英文書信這樣寫

Inquiry of Payment Term 詢問付款條件

Dear Tony,

We acknowledged the receipt of the quotation for your ball valve with many thanks. Before we place the order with you, we would like to know if it's acceptable to change the payment term to D/P at sight, as most of our suppliers dealing with us.

We look forward to your favorable reply to kick off the business relation benefiting to each other.

Sincerely yours,
Tom Smith

中文翻譯

湯尼 您好：

　　我們公司確認收到貴公司球閥報價，感激不盡。下訂單予貴公司之前，我們想要知道是否貴公司可接受付款條件改為即期付款交單，就如同大部份供應商與我們公司交易的方式。

　　期盼貴公司賜予佳音，以啟動雙方互利合作關係。

湯姆 史密斯 敬啟

▶▶ 信用狀雖然是國際貿易中的最普遍使用的付款條件，但信用狀並無統一的格式。即便如此，其主要內容基本上是相同的，大致上來說包含如下：

● 信用狀說明：信用狀的種類、性質、編號、金額、開狀日期、有效期及到期地點、當事人的名稱和地址、信用狀可否轉讓等。

● 匯票說明：出票人、付款人、匯票的期限以及出票條款 等。

● 貨物說明：貨物的名稱、品質、規格、數量、包裝、運輸標誌、單價等。

● 運輸要求：裝運期限、裝運港、目的港、運輸方式、運費付款方式，可否分批裝運和中途轉運 等。

● 單據要求：單據的種類、名稱、內容和份數等。

● 特殊條款：依據進口國政治、經濟、貿易情況，或依各別交易的差異，可做出不同的規定。

● 信用狀當事人及關係人的相關責任文句。

▶▶ 信用狀所載內容正確與否關係著未來是否可順利議付或準時提貨，因此需審慎核對內容，務必確保與相關商業文件內容一致。

翻譯小試身手

❶ 因為我們的預算極為有限,請您給予支援,提供特別折扣給我們。

❷ 面對愈來愈多的廠商爭奪各自的市場佔有率,我們免不了將會面臨激烈的競爭。

參考答案

MP3 26

❶ Our budget is extremely limited. Please support this project by offering a special discount to us.

❷ With more manufacturers vying for the market share, we'll unavoidably face intense competition.

14 要求修改信用狀 Request of L/C Amendment

 E-mail 這樣寫

❶ Please find attached your Order Acknowledgment for PO No. 1202. The Order Acknowledgment confirms the shipping date and related details. Kindly please look over this to ensure accuracy. Should you have any questions or queries, please do not hesitate to contact me.

請見您訂購單單號 1202 的訂貨確認單如附，此確認單上會確認出貨日期及其他明細，請逐一審閱，以確保正確無誤。若您有任何的問題，請儘管與我聯絡。

❷ Please find attached our Sales Acknowledgement in relation to your Purchase Order number referenced. Please check the Acknowledgement and in the event of any discrepancies, advise us immediately by return e-mail.

附上我們對您上述單號之訂購單的銷售確認單，請檢查一下此確認單，如有任何不符之處，還請立即以 e-mail 告訴我們。

英文書信這樣寫

Request to Amend L/C 要求修改信用狀

Dear Tom,

　　After reviewing to the L/C against your P/O No. 1234, we notice that there's any insufficiency in the amount. The amount of your L/C is only USD3,000, which is USD300 shorter than the total value of your order.

　　Please make the L/C amendment to increase the amount up to USD3300 at once to ensure the production and shipment of your order in time. Your prompt reply will be highly appreciated.

Sincerely yours,
Tony Yang

中文翻譯

湯姆 您好：

　　在檢視貴公司針對訂單號碼 1234 所開出的信用狀後，發現金額不足。貴公司信用帳金額為 3000 美元，比貴公司訂單金額短少 300 美元。

　　請立即修改信用狀增加金額至 3300 美元，以保證及時生產及裝運。貴公司儘速回覆，我們公司將感激不盡。

湯尼 楊 敬啟

▶▶ 銀行信用的運作過程

第一階段　國際貿易買賣雙方在貿易合同中約定採用信用狀付款。

第二階段　買方向所在地銀行申請開狀。開狀要交納一定數額的信用狀定金，或請第三方有資格的公司擔保。

第三階段　開狀銀行按申請書中的內容開出以賣方為收益人的信用狀，再通過賣方所在地的往來銀行將信用狀轉交給賣方；賣方接到信用狀後，經過核對信用狀與合同條款符合，確認信用狀合格後發貨。

第四階段　賣方在發貨後，取得貨物裝船的有關單據，可以按照信用狀規定，向所在地銀行辦理議付貨款。

第五階段　議付銀行核驗信用狀和有關單據合格後，按照匯票金額扣除利息和手續費，將貨款墊付給賣方。

第六階段　議付銀行將匯票和貨運單寄給開狀銀行收賬，開狀銀行收到匯票和有關單據後，通知買方付款。

第七階段　買方接到開狀銀行的通知後，向開狀銀行付款贖單。贖單是指向開狀銀行交付除預交開狀定金後的信用狀餘額貨款。

翻譯小試身手

❶ 我很高興我們雙方能夠達成妥協，我們會接受您前一封 e-mail 所提出的付款方式。

❷ 這間研究機構對此產業的發展有很大的影響力。

參考答案

MP3 28

❶ I am happy that we are able to reach a compromise. We will accept your proposal for payment as described in your last e-mail.

❷ The research institute has a substantial influence on the development of this industry.

15 通知銀行改信用狀 Notify Bank of L/C Amendment

 E-mail 這樣寫

 MP3 29

❶ We did place our official order to you but our customer's project is postponed to a later date and; therefore, we have to adjust the time scale and quantity of our order as follows. We appreciate if you could confirm your acceptance at your earliest convenience.

我們有下了正式訂單給您，但是我們客戶的案子往後延了，所以我們得調整我們訂單的時間排程和數量，請見明細如下，若您能盡快跟我們確認接受此修改，我們將會很感激。

❷ We'd like to decrease the quantity of our order if at all possible. This is due to the fact that our customer's experiment was delayed, so his exact demand at this stage will be lower than the ordered quantity. Please confirm the order change.

如果可以的話，我們想要減少我們所訂的數量，因為我們客戶的實驗有延遲，因此他現階段的實際需求會比訂購量來得少。還請您確認此訂單變更。

英文書信這樣寫

Request of second L/C Amendment 二次修改信用狀

Dear Tom,

We're sorry to inform you that we still haven't received the amended L/C No. LB 123 with correct amount as per our request. Besides, Owning to the delay amendment, we miss the original vessel and herein, must request to extend the date of shipment to June 1.

Your prompt attention to this matter will be greatly appreciated.

Sincerely yours,
Tony Yang

中文翻譯

湯姆 您好：

　　很遺憾通知貴公司，我們公司尚未收到依要求修改為正確金額的信用狀號碼 LB123。此外，茲因改狀延遲，導致我們公司錯過原訂的船班，在此，需要求將裝船日延期至六月一日。

　　貴公司若儘早處理此事，我們公司不勝感激。

湯尼 楊 敬啟

▶▶ 信用狀所附文件包含：

❶ 匯票 （Draft）	❷ 商業發票 （Commercial Invoice）
❸ 海運提單 （Marine / Ocean Bill of Lading）	❹ 保險文件 （Insurance Documents）
❺ 包裝單 （Packing List）	❻ 重量單 （Weight List/Certificate）
❼ 數量單 （Quantity Certificate）	❽ 產地證明書 （Certificate of Origin）
❾ 品質證明書 （Certificate of Quality）	❿ 受益人證明 （Beneficiary's Certificate）
⓫ 電報抄本（Cable Copy）	

▶▶ 實務上常見信用狀不符項目：

▶▶ **運輸相關：** 啟運港、目的港或轉運港與信用狀的規定不符、可否允許貨物短裝或超裝、裝運日期過期、沒有貨物裝船證明或註明「貨裝艙面」、運費由受益人承擔，但運輸單據上沒有「運費付訖」字樣 等。

▶▶ **匯票相關：** 匯票上面付款人的名稱及地址不符、匯票上面的出票日期不明等。

▶▶ **商業發票相關：** 發票上面的貨物描述與信用狀不符、發票的抬頭人的名稱、地址等與信用狀不符等。

▶▶ **包險單據相關：**保險金額不足、保險比例與信用狀不符、保險單據的簽發日期遲於運輸單據的簽發日期、投保的險種與信用狀不符等。

▶▶ **各種單據要求：**不潔運輸單據、各種單據的類別與信用狀不符、各種單據中的幣別不一致、匯票、發票或保險單據金額的大小寫不一致、匯票、運輸單據和保險單據的背書錯誤或應有但沒有背書、單據沒有簽字或有效印章等。

翻譯小試身手

❶ 若是客戶會一次下單訂購 20 組，那就能有更大折扣的協商空間。

❷ 對於這項產品在你們市場上的潛力，我們想要有更多的瞭解。

參考答案

MP3 30

❶ If the customer will be purchasing all 20 kits in one order, there may be room for negotiating a bigger discount.

❷ We'd like to understand in more detail the potential of this product in your market.

Unit 16

放寬付款條件 Request of Easing Payment Term

E-mail 這樣寫

❶ You can choose from express, sea and air freight for shipping your order depending on your budget and when you need your order to arrive.

您訂單的出貨方式有快遞、海運、空運可選擇，就看您的預算，以及您要您的訂單何時到貨了。

❷ We offer the following shipping methods:

- Airmail: Uninsured and untrackable, between 5-7 days delivery to Europe, longer to other parts of the world.

- UPS (various services): insured and trackable – delivery speed can be next day for some parts of the world if the proper service is chosen.

我們有提供下列的出貨方式：

一郵寄：沒有保險，無法追蹤，寄到歐洲約 5～7 天可到貨，寄到其他區域則需時更長些。—UPS（多種服務）：有保險，可追蹤，若選擇適當的服務類型，有些區域可在隔天就到貨。

英文書信這樣寫

Part
1
國貿基礎篇

Part
2
國貿進階篇

Notification of L/C Acceptance 信用狀承兌通知

Dear Tom,

We're writing to inform you that your P/O No. 1234 has been delivered on board on May 20 as schedule per the attached copies of the shipping documents.

In accordance with the irrevocable L/C No. LB 123 for USD 3,300 issued by London Bank, we have valued on you at sight against this shipment. The full set of shipping documents will be sent to you, once you accept the draft. Please kindly honor the draft immediately upon presentation.

Sincerely Yours,
Tony Yang

中文翻譯

湯姆 您好：

　　僅此文通知貴公司，號碼為 1234 的訂單已如期於五月二十日裝船，茲附上出貨文件附本。依據倫敦銀行開出之總額 3,300 美元的不可撤銷信用狀號碼 LB123，我們公司已對此批

　　貨開出見票即付之匯票。一經貴司承兌後，將寄出全套裝船文件予貴公司。懇請貴公司於見票後立即予以承兌。

湯尼 揚 敬啟

國貿經驗談

▶▶ 匯票（**Bill of Exchange**）I

- **Bill of Exchange（匯票）字樣**：匯票上應明確標明 Bill of Exchange（匯票）字樣。

- **匯票的出票條款（Drawing Clause）**

  出票條款又稱為出票依據。説明匯票是依據某個信用狀的指示而開發，及信用狀開證行將對匯票履行付款責任的法律依據。

▶▶ 匯票期限（**Tenor**）

- 匯票上必須明確表明是即期付款還是遠期付款。如果是即期匯票則為 "At Sight"，如果是遠期匯票，則應填寫遠期天數。例如 At 30 Days Sight，表示 30 天遠期。而匯票的期限一般有以下幾種：

- At sight 即期付款

- At (30, 60, 90, 180...) Day after Sight：見票後（30, 60, 90, 180...）天付款

- At (30, 60, 90, 180...) Days after Date of Issue：出票後（30, 60, 90, 180...）天付款

- At (30, 60, 90, 180...) Days after Date of Bill of Lading：提單的出單日期後（30, 60, 90, 180...）天付款

▶▶ 匯票的金額（**Amount**）

- 匯票上的金額必須有小寫和大寫兩種表示。金額為整數時，大寫金額未尾處必須加打"Only"字樣，以防塗改。小寫金額必須與大寫金額完全一致。匯票金額的幣種應和信用狀金額的幣種

完全一致。

▶▶ **利息條款（Interest Terms）**
- 如果信用狀中規定有匯票利息條款，則匯票上必須明確載明，匯票上的利息條款文句一般包括得率和計息起訖日期等內容。

翻譯小試身手

❶ 新價格將於下個月生效，因此，到這個月底前，您還能以目前的價格來訂購我們的產品。

❷ 繼九月時預先寄發 e-mail 通知之後，我們將在 2017 年一月一日調漲費率。

參考答案   MP3 32

❶ New prices will come into effect next month, so you have until the end of this month to purchase our products at their current prices.

❷ As a follow-up to the e-mail notice sent in September, we will be implementing a rate increase on January 1, 2017.

17 託收付款 Payment term as Collection

E-mail 這樣寫

MP3 33

❶ Please send the product back in the original box (or any similar insulated box). The product is temperature sensitive and must be shipped at the appropriate temperature. So please have it packed with frozen gels and a minimum of 30 lbs dry ice.

請以原出貨箱子（或任何類似的絕緣箱）將此產品退回給我們，因為此產品對溫度敏感，必須在適合的溫度下退回，因此，請與冷凍凝膠一起包裝，同時箱內至少要放 30 磅的乾冰。

❸ We put as many ice packs as we can in the EPS box. Please note that these items are perishable and must be temperature controlled. Any delay in customs could damage the product.

我們在保麗龍箱子裡盡可能地多放冰包，請注意這些產品易腐壞，必須有溫度控制，若通關有任何延遲發生，都會損壞產品。

英文書信這樣寫

Request of Acceptance 要求承兌匯票

Dear Tom,

This is notifying you that your P/O No. 1234 had been delivered on board on May 29 as schedule per the copy of shipping document as the attached. In the meantime, the draft against invoice amount USD 3,300 has been through the remitting bank on documentary collection.

Please make payment against our documentary draft the sooner you can. Thanks for your kind support to Best Corp.

Sincerely yours,
Tony Yang

 中文翻譯

湯姆 你好：

以此文通知您，貴公司第 1234 號訂單已如期在 5 月 29 日裝船，如附件裝船文件副本所示。

同時，依據發票金額 3,300 美元所開出的匯票已交由託收行按跟單託收。請儘速依跟單匯票付款。感謝貴公司對倍斯特公司的支持

湯尼 楊 敬啟

匯票（Bill of Exchange）II

▶▶ 匯票的抬頭（Payee）

匯票的抬頭人就是匯票的收款人，在信用狀業務中，匯票的抬頭人經常為信用狀的受益人或議付行。

▶▶ 出票日期和出票地點

匯票的出票日期不得遲於信用狀的有效日期，但也不得遲於信用狀的最後交單期。匯票的出票地點一般為出口公司的所在地。

▶▶ 匯票的付款人（Drawee）

在信用狀業務中，匯票的付款人一般為信用狀的償付行、付款行、承兌行、保兌行或開證行。如果信用狀中沒有明確規定匯票的付款人，則應視開證行為付款人。

▶▶ 匯票的出票人（Drawer）

在信用狀業務中，匯票的出票人一般是信用狀的受益人，即出口公司。

▶▶ 匯票的背書

如果匯票的抬頭是 Pay to the order of China Export Company （The Beneficiary）（付給信用狀的受益人的指定人）或 Pay to China Export Company (The Beneficiary) or order（付給信用狀的受益人或其指定人），而匯票是由信用狀的受益人出具的匯票

應由出票人作背書，而議付銀行寄單索匯時也作背書。

如果匯票的抬頭是 Pay to the order of ● BANK（付給***銀行的指定人）或 Pay to ***BANK or order（付給***銀行或其指定人），而匯票是由信用狀的受益人出具的，出票人不應作背書，而議付銀行寄單索匯時作背書。

翻譯小試身手

❶ 附上我們 2017 年的價格表給您，有一些變動，不過大多數的產品都還是維持一樣的價格。

❷ 為了贏得這個訂單，我們以特例處理，也大幅降低了利潤，因此我們要求您也比照辦理。

參考答案

MP3 34

❶ I have attached our 2017 distributor price list for you. There have been a few changes but the majority of products have remained the same price.

❷ As we have made an exception in order to win this order and significantly lowered our margin, we ask that you do the same.

18 Refuse of Acceptance
拒絕承兌

 E-mail 這樣寫

MP3 35

❶ Your order has been shipped earlier today and should arrive shortly. Please note the AWB no. is 610792120117. Per your request, attached please find the complete set of shipping documents. The original ones will be packed together with the shipment.

您的訂單在今天稍早時已出貨，應該很快就可到貨，請注意提單號碼為 610792120117。另依您所要求，我在此附上一份完整的出貨文件，正本將會隨貨寄出。

❷ Once your order has been processed, our dispatch team will contact you and send you the shipping documents, including Packing Slip, commercial Invoice and also the Invoice for Payment.

等處理了您的訂單之後，我們的出貨人員就會立即跟您連絡，寄送出貨文件給您，文件包括裝箱單、商業發票，以及供付款所用的發票。

英文書信這樣寫

Refuse of Acceptance 拒絕承兌

Dear Tony,

We're much surprised to receive bank's notification that you have drawn a bill of exchange on us for our P/O No. 5678. Per our notification by e-mail dated Apr. 20, the said P/O was cancelled due to your unpunctual shipment.

Please kindly understand that there's no reason for us to honor the draft.

Sincerely yours,
Tom Smith

 中文翻譯

湯尼 您好：

　　我們公司十分驚訝收到銀行通知貴公司針對我司第 5678 號訂單開出匯票。根據我司四月二十日信文通知，上述訂單因貴司無法準時出貨而取消。

　　請諒解我們公司拒絕承兌匯票。

湯姆 史密斯 敬啟

▶▶ 託收 Collection

指出口商向進口商開出匯票，透過本國託收銀行委託其國外往來銀行向進口商商收取票款。可分為**跟單匯票託收（Documentary Bill）**，即 D/P 付款交單及 D/A 承兌交單，以及**光票託收（Clean Bill）**：

❶ **跟單匯票（Documentary Bill of Exchange）；押匯匯票**：出口商開出商業匯票，將其對外國進口商的債權，轉讓予外匯銀行，並以出貨文件作為質押依據，以貼現方式取得債權。

❷ **光票（Clean Bill）；信用匯票**：出口商開出商業匯票，將其對國內進口商的債權，轉讓予外匯銀行，並「不」交附出貨文件，即兌收現款。

▶▶ 跟單匯票託收流程：

❶ 出口商與進口商簽署合約，同意依據跟單託收之方式收取款項。

❷ 出口商安排出貨並將出貨文件及相關託收單據送交託收銀行。銀行將託收指示送交進口商銀行（代收行）。

❸ 代收行通知進口商，由其決定是否付款。進口商支付款項或承兌匯票，並領取出貨文件。

❹ 代收行將款項匯至託收銀行。託收銀行將款項存入出口商帳戶。

▶▶ 匯款的方式：

❶ **T/T 電匯（Telegraghic Transfer）**：匯款人（即出口商），以定額本國貨幣交付予本國外匯銀行兌換成定額外匯，再由本國

外匯銀行以密碼電報方式通知國外受款人（即進口商）所在地之分行或代理銀行進行付款。

❷ **D/D 票匯（Demand Draft）**：匯款人（即出口商）以定額本國貨幣交付予本國外匯銀行兌換成定額外匯，並填具付款委託書，再由本國銀行將付款委託書寄送到國外受款人（即進口商）所在地之分行或代理銀行進行付款。

❸ **Travelers L/C 旅行匯信**：以定額本國貨幣交付予本國外匯銀行兌換成定額外匯，並申請開發旅行信用狀，即可依據此旅行匯信，在限額內發出匯票，請求開狀銀行的分行或代理行兌付款項。

翻譯小試身手

❶ 買方要負擔所有海外交易所產生的任何相關銀行費用。

❷ 因為生產成本高，我們無法降至跟競爭者相同的價格水準。

參考答案
 MP3 36

❶ The buyer is responsible for any associated bank charges on all overseas transactions.

❷ Due to high production costs, we are not able to match the competitor's pricing.

Unit 19

承兌延遲
Delay of Acceptance

 E-mail 這樣寫 MP3 37

❶ I've asked Singapore Airlines to make an amendment to the discussed AWB. Attached please find the Amended AWB.

我已要求新加坡航空修改我們討論的那份提單了,請見提單修改版如附。

❷ This shipment needs to go through document scrutiny at our customs. Since total 3 boxes were shipped, we're requested to submit the Packing List which contains box no. on it. Please revise your Packing List accordingly and e-mail to me asap.

這次的出貨得經過我們海關的文件審核程序,因為總共出了三箱,海關要求我們提供的裝箱單要列有箱號,還請您依照要求修改裝箱單,並請盡快 e-mail 給我。

英文書信這樣寫

Payment Term as T/T 電匯付款

Dear Tom,

We're pleased to advise you that the shipment of your P/O No. 1234 had been effected on May 29 as schedule. Attached please find the copy of shipping document.

We'll inform the forwarder to telex release the B/L, upon the receipt of the 70% balance against P/O amount by telegraphic transfer.

Shall there be any question, please feel free to let us know.

Sincerely yours,
Tony Yang

中文翻譯

湯姆 您好：

　　茲通知貴公司第 1234 號訂單已如期在 5 月 29 日出貨。請查收附件出貨文件副本。待收到貴公司電匯訂單金額的 70%餘款，我們公司將通知貨代電放提單。

　　如有任何問題，請不吝告知。

湯尼 楊 敬啟

商業發票（以下簡稱「發票」）沒有統一的格式，但是，一般應具備以下主要內容：

▶▶ 首文部分、發票名稱和號碼：首文部分應列明發票的名稱、發票號碼、合同號碼、發票的出票日期和地點，以及船名、裝運港、卸貨港、發貨人、收貨人等。發票上應有 Invoice（發票）字樣。一般情況下應按信用狀對發票的具體要求製做，例如，如果信用狀要求提供的是 Commercial Invoice（商業發票），則發票的名稱必須有 Commercial 字樣，否則發票則與信用狀的要求不符。發票號碼是出口商製作發票的編號，這是發票中不可缺少的內容之一。

▶▶ 合約號碼、日期和地點：發票中應反映進出口雙方貿易合同的號碼。信用狀中經常規定要示在發票上加註合同號碼。發票的日期就是發票的製作日期，一般情況下，發票的日期應略早於匯票日期，同時不能遲於信用狀的有效期，發票的日期應在運輸單的出單日期之前。發票的製作地點一般是出口公司的出單地點。

▶▶ 裝運港和目的港、運輸工具：一般發票上應列明貨物的裝運港和目的港口。如果採用直達船運輸時，應在發票上加註船名，如果中途需要轉船，則應註明轉運港名稱。

▶▶ 收貨人（Consignee）：收貨人就是發票的抬頭人，一般為進口商或信用狀的開狀申請人，其名稱和地址（有時包括電傳號碼、

傳真號碼等）應嚴格按照信用狀的具體要求加打在發票上。

▶▶ 發貨人（**Consignor**）：發貨人一般為發票的出票人，即是出口商或信用狀的受益人，其名稱和地址一般列印在發票的上方。一般情況下，發貨人的名稱和地址等應與信用狀的受益人完全一致。另外，發貨人（即信用狀的常駐益人）應在發票的右下方列印上自己的名稱並簽字或蓋章。

翻譯小試身手

❶ 公車與地鐵車資調漲今已生效，其價格調漲幅度比通貨膨脹率還大上三倍有餘。

❷ 以信用卡付款對我們來說需要額外的處理作業，因此，我們可能會要收取 USD 30.00 的手續費。

參考答案 MP3 38

❶ Bus and train fare increases have come into effect, with prices rising by more than three times the rate of inflation.

❷ Payments via credit card mean additional work for us and therefore we may need to charge a handling fee of USD 30.00.

 E-mail 這樣寫 MP3 39

❶ We noticed that your package has encountered a clearance delay and stuck in your customs. The customs need clearance instructions from the recipient so as to authorize release.

我們注意到出給您的包裹在通關上有延遲，卡在您們的海關，而海關需要收貨人的清關指示才能放行。

❷ For export orders, we always ask our customers to make payment in advance. But, as a gesture of goodwill, I could make an exception for you and accept "net 10 days" as an alternative way of payment.

對於出口訂單，我們都會要求客戶預付貨款，不過，為了表示善意，我可為您破個例，接受「淨 10 天」這個不同的付款條件。

英文書信這樣寫

Payment Term as T/T 電匯付款

Dear Tom,

We're pleased to advise you that the shipment of your P/O No. 1234 had been effected on May 29 as schedule. Attached please find the copy of shipping document.

We'll inform the forwarder to telex release the B/L, upon the receipt of the 70% balance against P/O amount by telegraphic transfer. Shall there be any question, please feel free to let us know.

Sincerely yours,
Tony Yang

中文翻譯

湯姆 您好：

　　茲通知貴公司第 1234 號訂已如期在 5 月 29 日出貨。請查收附件出貨文件副本。

　　待收到貴公司電匯單訂單金額的 70%餘款，我們公司將通知貨代

電放提單。

如有任何問題，請不吝告知。

敬啟者 湯尼 楊

國貿經驗談

本文部分
發票的本文主要包括嘜頭、商品名稱、貨物數量、規格、單價、總價、毛重/淨重等內容。

▶ 嘜頭（Shipping Mark）：嘜頭是貨物裝的標記。發票上的嘜頭應與信用狀中規定的嘜頭完全一致。

▶ 商品名稱：發票上的商品名稱應與信用狀中規定的商品名稱完全一致。如果信用狀中的商品名稱有錯誤或漏字等並且未來得及修改，發票上的商品名稱也應將錯就錯，以保證發票與信用狀規定的完全一致。不過，可在錯誤的名稱後面加註正確的名稱。

▶ 貨物數量和重量：發票中列明的貨物的數量（件、個、條等）和重量應與信用狀中貨物描述的數量和重量完全一致，並與提單等基本單據中貨物的數量和重量一致。

▶ 規格：貨物的品質規格是出口商交貨的標準，發票上一般應標明所裝貨物的品質規格情況。

▶▶ 商品的單價和總價：發票的單價一般包括計價單位、計價貨幣、價格條件和單價金額等內容。發票的單價應與信用狀中規定的貨物單價一致。

發票的總價不是貨物總金額、總價值，也就是貨物數量與貨物單價之積。發票的總金額不能超過信用狀的金額（除非信用狀另有規定）。

翻譯小試身手

❶ 我們會寄送 100 份免費型錄給我們的代理商，但運費得由各個代理商來負擔。

_____

_____

❷ 我們也可接受信用卡付款，這部分的交易會要加收 3 % 的處理費。

_____

_____

參考答案

❶ We'll send 100 free catalogs to our distributors, but each distributor is responsible for shipping expenses.

❷ Payment by credit card is also acceptable, but we may add a 3% surcharge for the processing fee for credit card transactions.

 E-mail 這樣寫 MP3 41

❶ If you are sending the payment using Letters of Credit, please make sure that your Letter of Credit has to be irrevocable, negotiable with our bank, issued or confirmed by a first class international bank.

若是您要用信用狀付款,請確定開不可撤銷、可讓購給我們銀行的信用狀,要由一級國際銀行來開立或保兌。

❷ We accept checks, credit cards, and wire transfers as payment methods. However, a $50 fee applies to all invoices paid via wire transfer due to high bank fees on our end.

我們可接受以支票、信用卡及匯款付款的方式,不過,若以匯款支付,因我們這邊銀行所收的銀行手續費高,故所有發票貨款須再加上$50。

英文書信這樣寫

Loss of Check 支票遺失

Dear Sirs,

With reference to our phone conversation today, whereby we ask you to stop paying the check No. 123. The above-mentioned check was drawn in settlement of our account with Best International Corp., but was missed during delivery. We'll write another check to take over.

Please pay your prompt attention to this matter and ensure to stop payment of check No. 123.

Sincerely yours,
Tom Smith

 中文翻譯

敬啟者

　　依據我們今日電話內容，在此要求您止付第 123 號支票。上述支票是為了支付我們公司對倍斯特國際公司的欠款，但支票在郵寄過程中遺失了。我們會重新開立新票取代。

　　請儘速處理此事，並確保止付第 123 號支票。

湯姆 史密斯 敬啟

國貿經驗談

電放提單（Telex Release B/L 或 Surrendered B/L）

▶▶ 是指船公司或其代理人簽發並註明 Surrendered 或 Telex Release 字樣的提單。簡而言之是指託運人（出口商/賣方）不用領取正本 提單（B/L），而收貨人（進口商/買方）無須憑正本提單，僅憑 蓋章的電放提單傳真或身份證明即可提取貨物的一種放貨形式。

▶▶ 其申請流程為託運人（出口商/賣方）填寫電放申請書並提供保函 後傳真給船公司，再由船公司提供「保函」及電放信函通知目的 港代理無須憑正本提單放貨。保函的內容通常有，託運人名稱 （shipper's name）、航班（voyage No.）、啟航日（Sailing date）、提單號碼（B/L No.）及貨代無條件免責條款。而電放申 請書上經常批註字句 Please kindly release cargo to Consignee here - below without presentation of the original 承運人名稱或 或代名稱 Bill of Lading，意指在不提示承運人/貨代的正本提單 下，請將此批貨發放給以下所載之收貨人。

▶▶ 電放通常以 PP（預付運費，Prepaid）為付款條件。此外有下列 情形者不可電放：

● 以信用狀（L/C）為付款條件時。

● 以託收（Collection）為付款條件時。

● 提單（B/L）上收貨人（Consignee）欄位未填寫收貨人全名 時。此其況通稱為「order 單」，即收貨人欄位出現 order 字

樣，例如：Order 或 Order of ABC Co.

● 空白提單（B/L）時。即提單（B/L）中收貨人（Consignee）
欄位空白。

翻譯小試身手

❶ 我們可以配合您的要求，提供這一項產品的訂製包裝。

_____

_____

❷ 為了能與其他廠牌競爭，我們希望您能同意提供 50%的折扣。

_____

_____

參考答案

MP3 42

❶ We're able to accommodate your request to supply the custom
size of this product.

❷ To compete with other brands, we hope you will approve a 50%
discount.

 E-mail 這樣寫 MP3 43

❶ For credit card transactions, we'll surcharge additional 3% as processing fee. Also, please fill out the attached credit card authorization form and send it back to us.

我們對信用卡交易會另收 **3%**的處理費，另也請填寫所附的信用卡授權書，填完後再發回給我們。

❷ Our records indicate the attached invoice is past due. Please acknowledge receipt of this email immediately and arrange for prompt payment of this overdue amount.

我們的紀錄顯示附件的發票已逾期了，請立即回覆您有收到此 email，也請盡速支付此逾期貨款金額。

Part
1
國貿基礎篇

Part
2
國貿進階篇

 **英文書信這樣寫**

Enquiry to Insurance Risks (from Buyer) 洽詢保險條件（買方詢問）

Dear Tony,

Just a quick note to confirm that we have received your quotation on CIF basis per email dated May 15. Before placing the order, we'd like to know the coverage of your insurance. Please share with us details for our better understanding.

We hope to hear from you soon.

Sincerely yours,
Tom Smith

 中文翻譯

湯尼 您好：

只是簡單的確認我們公司已收到貴公司五月十五日以信函提供的運保費在內報價。在下單前，我們公司想要瞭解貴公司提供的保險含蓋的範圍。請提供明細，以利我們公司瞭解。

靜候佳音。

湯尼 史密斯 敬啟

匯票、本票、支票的區別

▶▶ 匯票、本票、支票同屬狹義的票據範疇，其構成要素大致相同，都具有出票、背書、付款這些流通證券的基本條件，都是可以轉讓的流通工具。它們之間的主要區別是：

▶▶ 匯票和支票有三個基本當事人，即出票人、付款人、收款人；而本票只有出票人（付款人和出票人為同一個人）和收款人兩個基本當事人。

▶▶ 支票的出票人與付款人之間必須先有資金關係，才能簽發支票；匯票的出票人與付款人之間不必先有資金關係；本票的出票人與付款人為同一個人，不存在所謂的資金關係。

▶▶ 3 支票和本票的主債務人是出票人，而匯票的主債務人，在承兌前是出票人，在承兌後是承兌人。

▶▶ 遠期匯票需要承兌，支票一般為即期無需承兌，本票也無需承兌。匯票的出票人擔保承兌付款，若另有承兌人，由承兌人擔保付款；支票出票人擔保支票付款；本票的出票人自負付款責任。

▶▶ 支票、本票持有人只對出票人有追索權，而匯票持有人在票據的效期內，對出票人、背書人、承兌人都有追索權。匯票有複本，而本票、支票則沒有。支票、本票沒有拒絕承兌證書，而匯票則有。

Part
1
國貿基礎篇

Part
2
國貿進階篇

翻譯小試身手

❶ 為了要給個具競爭力的價格，我們大幅調降了利潤，因此，我們要求您們公司也要這麼做。

❷ 我們的最終使用者價格加計了 **15**％的固定加成。

參考答案

MP3 44

❶ We're significantly decreasing our margin in order to provide competitive pricing; therefore, we ask that your company does the same.

❷ Our end user pricing includes a fixed markup of 15%.

 E-mail 這樣寫 MP3 45

❶ Please settle all the accounts immediately as they are substantially overdue. Please acknowledge receipt of this notice and indicate the date that the payment will be made.

請立即付清所有款項，因已逾期多時，請您回覆有收到此通知，並告知將於哪一天付款。

❷ Please be reminded that Interest will be payable on all accounts due and outstanding for over 30 days at the rate of 2% per month from the due date of invoice to the date of payment.

在此提醒您，所有逾期 30 天仍未付的貨款將以月息 2 厘計算利息，所計期間由發票到期日起，計至付款日為止。

英文書信這樣寫

Request of Additional Insurance (from Buyer) 請求額外保險
（買方需求）

Dear Tony,

With reference to our P/O No. 2233 for 1,000pcs of ball valve placed on C&F basis, please arrange ICC (A) insurance for us at 10% of the total invoice value, which is USD5,000. The premium will be paid by us upon the receipt of the relative insurance documents from you. If you have any further questions, please feel free to let me know.

Sincerely yours,
Tom Smith

中文翻譯

湯尼 您好：

　　關於我們公司第 2233 號訂單，依運費在內條件訂購貴公司 1,000 件球閥，請協助替我們公司依據商業發票總額 10%的費用：金額 5,000 美元，總額 50,000 美元，投保協會貨物 A 款險。待收到貴公司提供的相關保險憑證後，我們公司將會支付保費。如有任何進一步問題，請不吝告知。

湯姆 史密斯 敬啟

國貿經驗談

▶▶ 進出口貨物運輸保險費率是自由費率，其計價基礎常因下列因素而有所調整：

- 貨品名稱及其種類，特殊貨物風險比較高，費率相對較高
- 目的港狀況，如目的地的治安狀況等。一般來說，歐美線費率較低，而非洲南美費率較高
- 承保險別及其它承保條件，如加保戰爭、罷工、民變等險種，費率較高
- 運輸工具狀況，如船名、船齡、船級、船籍等，以確定是否承保及費率高低
- 運輸方式，如為整櫃運輸或散裝運輸，是否整船運輸
- 包裝方式，如為木條箱包裝或裸裝
- 要保公司的業務量
- 要保公司的出險記錄
- 其它

▶▶ 保費計算：
保費＝保險金額 × 保險費率
保險金額＝發票金額／信用狀金額 × 110%

▶▶ 通常每筆保單的最低保費為 400 元。

▶▶ 實務上投保行為需在運送風險發生前辦理，因進出口業務之不同而區別如下：

▶▶ 出口業務：應在向船公司/航空公司簽訂裝貨單（S/O）及辦理報關時，同時申辦保險。可先將要保書送交保險人，而後再補齊相關資料。

▶▶ 進口業務：進口商應在申請輸入許可證或開發信用狀時，同時申辦保險，務必確保在啟航前投保，而保險人不負責在保險單簽發前所發生之任何損失賠償。

翻譯小試身手

❶ 當我們提供更大折扣時，我們希望您也能同樣調降利潤，以爭取客戶。

❷ 請提供更多您公司背景的資訊，例如人數、銷售額與成長率。

參考答案 MP3 46

❶ When we issue larger discounts, we expect you to reduce the profit also to win customers.

❷ Please provide more background information about your company, such as your headcount, sales revenue, and growth rate.

 E-mail 這樣寫 MP3 47

❶ I have attached a statement of your account as of today. Please check your records to make sure that you are not missing any Invoices / Credit Memos.

我在此附上您公司到今天的對帳單,請查查您的紀錄,確認一下是否沒有漏失所列的任何發票/貸項通知單。

❸ Our auditors are auditing our financial statement and wish to obtain direct confirmation of amount shown below as of Dec. 31, 2016. Please compare with your records and note the details of differences, if any.

我們的稽核人員正在稽核財務對帳單,希望可取得您的直接回覆,確認下面列至 2016 年 12 月 31 日的金額,請與您的紀錄比較,並標記出任何有差異的明細資料。

 **英文書信這樣寫**

Request to Make Claim on Sb.'s Behalf 請求代理索償

Dear Tony,

This is to inform you the shipment via S.S "Ever Lmbent" has already arrived at London, but noticed that there is a severe cosmetic defect by insurance surveyors.

As the insurance was covered at your side, please take up the claim for us with the insurance company upon the receipt of the surveyor's report and Broken and Damaged Cargo List. Appreciate your kind assistance.

Sincerely yours,
Tom Smith

中文翻譯

湯尼 您好：

　　茲此文通知貴公司，由 "Ever Lmbent" 號承運的貨物已運抵倫敦，但保險驗貨員發現貨物有嚴重的外觀不良。

　　由於該保險由貴公司承辦，待受到檢驗報告及貨物殘損單時，請協助我方向保險公司提出索賠感謝貴公司的協助。

湯姆 史密斯 敬啟

國貿經驗談

海上貨物運輸保險的索賠時效：

▶▶ 依相關國際公約和各國法律規定為基準，在須執行海上貨物索賠之情形時，收貨人或索賠人應在規定的時間內向承運人及時發出貨運事故書面通知書，聲明保留貨運事故索賠權。

▶▶ 承運人交付貨物予收貨人時，通常收貨人與承運人會對貨物進行聯合檢查，若收貨人未及時以書面方式，通知承運人有關於貨物損壞之情事者，則視為交付之貨物狀況良好，以此為證。而後相關損壞賠償之舉證責任就由承運人轉移至收貨人，如果收貨人之後無法舉證，證明貨物損壞屬承運人之過失，則將無法順利進行後續索賠。

▶▶ 依《海牙規則》，承運人負舉證責任，如果索賠人拖延通知，將不利於承運人舉證。但是否作出事故通知並不影響其索賠權利，亦即如果收貨人能夠舉證證明貨運事故是由承運人所造成，承運人仍應負損害賠償。反之，即使發出貨運事故通知書，而承運人能夠舉證證明貨運事故非其本身之責任，經確認後則承運人無需對貨運事故負損害賠償責任。

翻譯小試身手

❶ 您可以點入網站上的產品圖片，然後往下拉，就可看到完整的產品名稱。

_____

_____

❷ 總金額是要支付的最終價格。

_____

_____

參考答案

MP3 48

❶ You can click the product image on our website and then scroll down for the full product description.

❷ The grand total is the final price that must be paid.

Unit

25 洽詢船班 Enquiry of Shipping Schedule

 E-mail 這樣寫

 MP3 49

❶ I have included an updated copy of our product list with the new products added in August 2016. Additionally, I have attached two of our most recent flyers. Feel free to browse all our flyers at our website.

我有附上了我們產品表的更新版本，裡頭有 2016 年八月新增的產品。另外，附件還有我們兩份最新的單張型錄，也請上我們的網站，裡頭收有所有的單張型錄供您瀏覽。

❷ I've attached the manual for your review. Also below is a link to the Certificate of Analysis for our latest lot. Please note that the actual lot received may be different depending on lot availability at the time of order placement.

我在此附上產品手冊供您詳閱，也貼上我們最新批次的分析報告連結，請注意您實際收到的貨品批次可能會有所不同，因會視您下單當時的現貨批次來決定。

英文書信這樣寫

Enquiry of Shipping Cost by Exporter 洽詢運費（出口商詢問）

Dear Sirs,

We have received an order for 200 Metric tons of brass, which should be shipped in bulk during this month from HK to Port of London. Please provide your best freight rate and sailing schedule.

Please respond at your earliest convenience.

Sincerely yours,
Tony Yang

 中文翻譯

敬啟者：

我們公司收到一張 200 噸銅的訂單，並預計在六月份安排散裝由香港運送至倫敦港。請提供貴公司最優惠費率及船期表。

請盡速回覆。

湯尼 楊 敬啟

Part 1 國貿基礎篇

Part 2 國貿進階篇

▶▶ 定程租船的類別：

● 單程租船（Single Voyage Charter）：又稱「單航次租船」，即只裝運一個航次，該次航程結束時，租船合約即終止。

● 來回程租船（Round Trip Charter）：在完成一個航次後，接著再裝運一批貨返回。

● 連續單程租船（Consecutive Trip Charter）：在同一去向的航線上連續完成多個單航次運輸，船舶必須是去程運貨，回程空放，船舶所有人不能利用空船攬載其他貨物。

● 包運合約租船（Contract of Affreightment）：船舶所有人在約定的期限內，派遣數艘船舶，將指定的批貨，由指定出貨港運送到指定目的港，不論其航程次數。

▶▶ 「船務公司」是指獨立營運且有自己船隻的公司，如長榮海運，陽明海運、或萬海海運。「船務代理公司」沒有自己的船隻，但代理多家船公司的攬貨業務。實務上，船務代理能提供更多船舶資訊及服務範圍也較廣。

翻譯小試身手

❶ 下單時只要加註促銷碼 0925PA，就可得到一件免費的 T 恤。

_____

_____

❷ 我們工廠搬遷並不會在製程上或在已建立的產品規格上有任何的變動。

_____

_____

參考答案 MP3 50

❶ Simply reference promo code: # 0925PA when placing your order and you'll get a free T-shirt.

❷ There will be no changes to the manufacturing processes or established product specifications as a result of the relocation of our factory.

 E-mail 這樣寫　　　　　　　　　　　　　　　　 MP3 51

❶ Free samples are available upon the request of end users. They must fill out their lab information on the sample request form online.

免費樣品只能提供給最終使用者，他們必須在線上的樣品要求表格裡，填上他們實驗室的相關資料。

❷ We're willing to send you a sample free of charge. Do you have a courier account? We could send a sample to you with freight collect.

我們願意免費提供樣品給您，請問您有快遞帳號嗎？我們可以用運費到付的方式寄樣品給您。

英文書信這樣寫

Shipping Instruction 裝船指示

Dear Tony,

We are delighted to inform you that we have opened an irrevocable sight L/C No.125 with amount USD20,000 in your favor. Please arrange the shipment as soon as you receive the same. We'd like to call your attention that each part should be wrapped with double cotton papers to avoid damage during transportation and packed in cardboard carton. Besides, the shipping mark should be as below:

ABC

P/O No. 2468   Nos. 3-20

London via H.K.   20 x 10 x 8 FT.

G.W. 500 KGS   Made In Taiwan

We'll await your shipping advice in soon.

Sincerely yours,

Tom Smith

湯尼 您好：

　　很高興通知貴公司我方已開立以貴公司為抬頭的不可撤銷即期信用狀，金額為 20,000 美元。 請收到後立即安排船運。請貴公司注意單一產品需使用雙層棉紙包裝，避免在運輸過程中受損，並使用硬紙箱包裝。此外，請依下列所示刷嘜：

ABC
P/O No. 2468　Nos. 3-20
London via H.K.　20 × 10 × 8 FT.
G.W. 500 KGS　Made In Taiwan

　　期待儘早收到貴公司的裝船通知。

湯尼 揚 敬啟

 國貿經驗談

▶▶ **裝船指示（Shipping Instruction）**、**裝船通知**：由進口商在貨物備妥之前發給出口商，載明所需貨物的裝運說明，通常包括船公司資料，裝船日期，開船日期，起運港口，卸貨港口，包裝需求等，其目的在於讓出口商有足夠的時間做好裝船準備。則是由出口商在貨物備妥裝船後發給進口商，通知貨物已裝船的通知單，通常包括貨物詳細裝運情況，其目的在於讓進口商做好付款和接貨的準備。

▶▶ **出口貨物明細單、裝貨單**：又稱「貨物出運分析單」，是指托運人依據買賣合約條款和信用狀條款內容所填寫，藉以向運輸公司或貨運代理公司申辦貨物托運的單證。又稱「關單」，是指承運人（船公司）在接受托運人提出托運申請後，發給托運人的單證，同時也是船長將貨物裝船的憑證。通常只有經海關簽章後的裝貨單，船方才能收貨裝船。

翻譯小試身手

❶ 這個產品有三種規格可供應，有提供量大的折扣。

❷ 謝謝您追加的品項，這些品項都已加進您最初的訂單了。

參考答案

MP3 52

❶ This product is available in 3 sizes, with bulk discounts available.

❷ Thank you for your added items. These have been included in your initial order.

 E-mail 這樣寫 MP3 53

❶ Have you been able to test the sample that we sent recently to you? We'd like to hear any feedback you have.

請問您已測試了我們最近寄給您的樣品了嗎？我們想知道一下任何您對此樣品的意見。

❷ The U.S. Department of Commerce has our following products on the Export Control List. An export license must be obtained prior to transferring the material to an international destination. If you are interested in purchasing these products, please complete the attached Letter of Assurance Application. It takes approximately 4-6 weeks for the license to be issued.

美國商業部將我們下列產品列入出口管制清單中，這些貨品在出口前，須先取得出口證。若是您有興趣購買這些產品，請填寫附件的確認書申請表，此出口證核發約需 4～6 週。

英文書信這樣寫

Delay Shipment (Complaint from Buyer) 交貨延遲（買方抱怨）

Dear Tony,

We have to tell you that the goods against our P/O No. 8888 were scheduled to arrive this Monday. However, we still haven't received the shipment till now.

We hope that you can find out the reason for the delay as soon as possible. Just a kind reminder that the delivery postponement for more than one week will break the terms of the contract.

Your prompt attention to this matter will be appreciated.

Sincerely yours,
Tom Smith

中文翻譯

湯尼 您好：

必須跟您說明，我們公司第 8888 號訂單的產品應於本週一到達，但截至目前我們公司仍未收到貨物。

我們公司希望貴公司能儘快查明運送延遲之原因。善意的提醒貴公司，若此批貨延期超過一週，將違反合約條款。

貴公司對此事的立即處理，我們公司將不勝感激。

湯姆 史密斯 敬啟

 國貿經驗談

▶▶ 漏裝 - 是指貨物已進港並完成報關，但在裝船時因故不能上船，則須由船公司安排裝載至另一艘船或延至下一航次，這種情形下，一般而言出口商不需再次報關。倘若海關不同意以（漏裝）為之，出口商則不得不辦理退關手續，待重新訂艙，並取得提單號碼後，再次做報關申報。

▶▶ 實務中，會面臨"拋櫃"情況大致分為兩類：

● 因出口商/製造商疏失，導致未能及時將貨櫃裝載上船，因而被「拋櫃」，此種情況下責任歸屬出口商/製造商。需註銷前次報關，並安排重新報關，稱為「改配」，所有相關費用由出口商/製造商負責。

● 艙位已滿或超載，因此被「拋櫃」，此種情況下責任歸屬船公司，稱為「漏裝」，通常貨櫃並未辦理退關。

翻譯小試身手

❶ 請簽名回傳訂單副本，以表示收到並接受此訂單。

_____

_____

❷ 請您準備好之後，開立正式的訂購單給我們，我們會立即處理。

_____

_____

參考答案 MP3 54

❶ Please acknowledge receipt and acceptance of this order by returning the copy duly signed.

❷ Please place your formal PO when you are ready and we will process it immediately.

Unit

28 數量不符 Error Shipping Quantity

 E-mail 這樣寫　　　　　　　　　　　　MP3 55

❶ If you find that your goods are faulty on arrival, then you are entitled to a repair, a replacement or a refund.

若您發現貨品到貨有瑕疵，您有權要求修理、補出貨或退款。

❷ If you discover that your goods are visibly damaged on arrival, please contact us within 7 days with details of the damage. We'll then arrange to ship a replacement to you.

若您收到貨後發現有明顯損壞的情況，請在七天內跟我們聯絡，並請提供損壞的詳細資訊，然後我們就會安排寄出替換的貨給您。

 英文書信這樣寫

Cosmetic Defect (Notification from Buyer) 外觀損壞（買方通知）

Dear Tony,

 I regret to inform you that nearly 25% of the goods against our P/O. 8888 delivered to us was received in handling damage as shown on the attached photos. I'll send the defective parts back to you for exchange.

 Please let us know the process regarding to receive replacements and confirm when you receive the returned goods.

Sincerely yours,
Tom Smith

 中文翻譯

湯尼 您好：

 很遺憾通知貴公司，貴公司寄給我們公司第 8888 號訂單的產品有將近 25%有處理不當造成的損傷，如附件照片所示。我將寄還損壞的產品回貴公司，以利更換。

請告知我司有關替換貨物流程，待貴司收到退還貨物後，亦請通知我司知悉。

湯姆 史密斯 敬啟

國貿經驗談

▶▶ **不可抗力（Force Majeure）**，又稱「人力不可抗拒」，是指在合約簽訂後，非因契約雙方當事人任何一方之過失或疏忽，而是發生了當事人無法預見且無法事先防範的意外事故，以致於無法履行合約或不能如期履行合約。在此情況下，遭受意外事故的一方可以免除履行合約的責任或延期履行合約。

▶▶ 不可抗力的範圍通常可分為兩種情況：
- 自然災害：如 水災、火災、冰災、暴風雨、大雪、地震等。
- 社會因素：如 戰爭、罷工、抗爭、政府禁令等。

▶▶ 無論是自然災害或社會因素，不可抗力必須同時具備下述條件：
- 不可預見性：契約雙方當事人任何一方無法預見不可抗力事件的發生。例如 啟航前氣象報告已預告即將有風暴，但船長未聽廣播天氣預報即開船，結果遇上風暴造成貨物受損。雖暴風屬不可抗力，但此乃屬船長可預計。此種情況下仍需承擔責任。
- 不可避免性：即使出現了不可預見的災害，如果造成的後果是可以避免的，那麼也不構成不可抗力。例如 船舶在海上遇到風暴，可進附近避風港卻未進，因而導致貨物受損。此種情況下仍需承擔責任。
- 不可克服性：契約雙方當事人任何一方對該事件的後果無法加以克服，無法加以阻止。此種情況屬不可抗力的延伸。

翻譯小試身手

❶ 在付款方面，我們偏好的方式為銀行匯款，發票上有列出我們的銀行資料。

❷ 對於星期四之後才收到的任何訂單，都只能等到下一週才能處理及出貨。

參考答案 MP3 56

❶ To make payments, by our preferred method of bank transfer, our bank details are included on our invoices.

❷ Any orders received after Thursday will be processed and shipped the following week.

Unit
29
功能瑕疵
Function Defect

❶ In the event that any of our products are faulty, damaged on arrival or incompatible, we require you to return the goods to us for examination, after which a refund or a replacement will be issued.

若有到貨瑕疵、損壞，或是不相容的問題，請您退回貨品供我們檢測，檢測後我們就會核發退款或寄出替換品項。

❷ We shall repair or replace the defective product covered by the warranty and warrant the products free of defect in material and workmanship under normal use during the warranty period.

對於保固範圍裡發現有瑕疵的產品，我們會提供修理與換貨的服務，並於保固期間內正常使用的條件之下，保證產品在材料與工藝技術上不會發生故障。

英文書信這樣寫

Delayed Response (Complaint from Buyer) 回覆延遲（買方抱怨）

Dear Tony,

I'm writing about the production schedule of our open orders.

Several emails were sent to you to follow the updated status, but there has been no response from your end so far. That leaves us feeling very disappointed.

We urgently request a reasonable explanation from you about this issue.

Sincerely yours,
Tom Smith

 中文翻譯

湯尼 您好：

此文是關於我們公司在手訂單生產計畫事宜。

我們公司已多次發文詢問最新情況，截至目前為止仍未收到貴公司回覆。此情況令我們公司極感失望。

針對此事，急切需求貴公司提出合理說明。

湯姆 史密斯 敬啟

 國貿經驗談

▶▶ **糾正行動（Corrective Action）**，指為使預期績效與計劃重新恢復一致，而採取的措施。糾正行動可在各項控制過程保證有效的項目管理。有效的控制系統作用在於找出項目偏差、提出修正方案、達到預防偏差再發的目的。

▶▶ 糾正行動的類型可分為「立即糾正行動」（Immediate Corrective Action）和「徹底糾正行動」（Immediate Corrective Action）兩種：

● 立即糾正行動：是指立即將出現問題的項目矯正到正確的軌道上。

● 徹底糾正行動：先釐清項目偏差產生的原因（root cause），再從產生偏差的地方開始進行糾正。

▶▶ 品質客訴為買賣交易中最棘手的項目之一，嚴重者可能會影響單一客戶後續產品出貨，甚至平行擴及到同一生產製程中的其他客戶產品。因此多數的品質管理者為了能在最短時間內看見問題解決績效，大多採取直接糾正行動，常常只能治標不治本。然而採取徹底糾正行動，對項目偏差進行仔細分析、提出永久性的改善方案，對確實糾正工作績效與標準之間的偏差是非常有益的，最終才能達到有效的管理。

翻譯小試身手

❶ 若您無法取得我們網站上的資源訊息，請隨時與我聯絡。

❷ 我們邀請您參加我們的網路研討會，還請盡早登記，以免向隅。

參考答案

MP3 58

❶ If you have any difficulty accessing our website resources, please feel free to contact me.

❷ You're invited to attend our web seminar. Please register soon before it's too late.

Unit

30 回覆延遲
Delay Response

 E-mail 這樣寫

❶ The warranty does not apply if the faults were caused by negligent handling of the product, improper use together with products that were not produced by us, modified use or use other than in accordance with the operating instructions.

若因如下行為而導致故障,則不適用保固服務,這些行為包括產品操作上有疏忽的情況、搭配其他非我們生產產品的不當使用、改裝使用,或是其他不符操作指示規定的使用情形。

❷ Please note that to be eligible for return, items must be in their original purchase condition, include all product documentation, and be shipped within 30 days.

請注意,若要合乎退貨規定,產品須以其原購買時的狀態退回,包括所有產品文件,且要在 30 天內寄出。

英文書信這樣寫

Statement Error (Notification from Buyer) 結單錯誤（買方通知）

Dear Tony,

I think you may have made a mistake calculating the bill for our P/O No. 5566. According to our agreement, 50% discount would be granted for the cosmetic defective parts, but the full amount is still on our statement.

Please re-calculate and send us an amended bill soon. I look forward to your prompt response.

Sincerely yours,
Tom Smith

中文翻譯

湯尼 您好：

　　我想貴公司在結算我司第 5566 號訂單時發生錯誤。依據雙方同意，貴公司將對外觀瑕疵品給予半價，但卻仍然全額記帳到我司的帳單裡。

　　請重新核算後儘快把修改後的帳單寄予我們公司。期待貴公司儘早回覆。

湯姆 史密斯 敬啟

國貿經驗談

▶▶ **生產計劃（Production Schedule）**是指為同時滿足客戶對訂單的三大訴求「交期、品質、成本」，以及確保企業能獲得適當利益對生產的三大訴求「材料、人員、機器設備」所做的計畫，以確切準備、分配及使用的計劃。

▶▶ 執行生產計劃的最終目的：

- 確保交貨日期
- 穩定生產量，避免產能過剩或產能吃緊
- 控制庫存水平
- 原物料採購的參考依據
- 現場員工調度或職工招聘的參考依據
- 擴充生產設備的參考依據

▶▶ 生產計劃的內容一般包含如下：

- 產品名稱及零件名稱
- 生產的數量或重量
- 生產部門及單位（依個別產品生產工序不同而定）
- 起始日
- 生產需求完成日及交期

Part
1
國貿基礎篇

Part
2
國貿進階篇

翻譯小試身手

❶ 若您要修改所選內容，您只要點進「購物籃」，就可移除或修改所選產品的數量。

❷ 我們延遲出貨的原因是因為我們本來應該要做出貨系統維護，但取消了，所以我們今天就可出貨給您們。

參考答案

 MP3 60

❶ Should you wish to amend your selection, simply click on the 'Basket' where you will have the opportunity to remove or amend the quantity of the product selected.

❷ The reason that we delayed the delivery was because we were supposed to have maintenance on our shipping system. But, that was cancelled, so we can ship it for you today instead.

 E-mail 這樣寫 MP3 61

❶ If your customer doesn't want to keep the product, then you could ask them to return it to us, given that it has been properly stored this entire time.

若是您的客戶不想留著這項產品,那麼您可以退給我們,不過先決條件是在這段期間裡,產品都有適當儲存著。

❷ Products may not be returned except with our permission, and then only in strict compliance with our return shipment instructions. Any returned items will be subject to a 15 percent restocking fee.

貨品若未經我們許可則不得退回,且須嚴格符合我們的退運指示。任何退回的貨品皆須支付 15％的重新上架費。

英文書信這樣寫

Claim for Shipping Damage with Insurance Company
天然災害索賠（保險公司賠償）

Dear Tony,

We were just informed that the container containing our P/O No. 2233 was lost in transit due to bad weather. As such loss is covered by the insurance, we will make the claim with the insurance company.

However, as we need the missing lots urgently, it will be highly appreciated if you could arrange immediate production run for us and confirm the earliest delivery date.

Thanks in advance for your prompt attention on this matter.

Sincerely yours,
Tom Smith

中文翻譯

湯尼 您好：

我們公司剛接獲通知，承載我們公司第 2233 號訂單的貨櫃因氣象惡劣在運輸途中遺失了。

因此項損失屬保險承保的範圍，我們會向保險公司提出索賠。

然而，由於我們公司緊急需要此批遺失的貨物，如貴公司能安排立即投產並回覆最快交期，我們公司將不勝感激。

預先感謝貴公司立即處理此事。

湯姆 史密斯 敬啟

 國貿經驗談

▶▶ 簡單來說，國際貿易的結算是以物品交易、錢貨兩清為概念的有形貿易結算。而結算方式就如付款條件（Payment Term）章節所陳述，主要是以信用狀（L/C）、匯付（T/T）或託收（Collection）等結算方式。無論採用何種結算方式，都應特別注意所會面臨的風險及如何防範，例如支票/匯票的真假，使用匯票者是否為銀行開立帳戶的法人 等，預防出貨後收不到貨款，而造成貨款兩空的情形，對此萬萬不能掉以輕心！

翻譯小試身手

❶ 唯一有關的其他費用就是當您要取回貨物時所要負擔的關稅了。

❷ 請注意，基於安全考量，我們要求客戶不要用 e-mail 或傳真來發送信用卡明細。

參考答案

 MP3 62

❶ The only other fees associated would be the customs fees which you are responsible for when retrieving the product.

❷ Please be aware that due to security concerns, we ask customers not to send credit card details via e-mail or fax.

E-mail 這樣寫

MP3 63

❶ All returns must be authorized by an RMA Number. Please email your request for authorization number before shipping back any merchandise.

所有退貨皆須有 RMA 退貨授權號碼，在您退回任何貨給我們之前，請寫 email 給我們，要求授權號碼。

❷ When processing the return shipment, please make sure to indicate the following in the comments section of the Commercial Invoice:

This product was originally shipped to Taiwan via FedEx AWB no. 805726811202.

No sale / transaction has occurred. The value stated is for Customs purposes only.

當辦理退貨時，請確實在商業發票的備註欄裡註明下列事項 此產品原出口至台灣的 FedEx 提單號碼為 805726811202 無銷售／交易發生。所列的金額僅供海關參考之用。

Part 1 國貿基礎篇

Part 2 國貿進階篇

英文書信這樣寫

Claim for Quantity Shortage 數量短出索賠

Dear Tony,

We received the goods against our P/O No. 2255 that you sent to us, but found one carton wan't full. We think it might have been caused by careless packing.

Please advise if you prefer to make replacement or refund for the shortage, said 2pcs.

Sincerely yours,
Tom Smith

中文翻譯

湯尼 你好：

收到貴公司寄送予我們公司的第 2255 號訂單的貨物，但發現其中一箱未滿箱。我們認為這是由於包裝疏失所致。

請告知貴公司傾向補貨兩件不足數，或退款。

湯姆 史密斯 敬啟

國貿經驗談

▶▶ 在不同貿易條件下，進口商對貨物投保的起迄點也有所差異，就
一般實務上常使用的貿易條件簡述如下：

- **工廠交貨（EX Works）**：應安排出口國之內陸運輸至買方倉庫
之貨物運輸保險。

- **船上交貨（FOB）**：應安排自出貨港至買方倉庫之貨物運輸保
險。

- **運保費在內（CIF）**：此項貿易條件已包含保險，但需特別注意
如承保範圍已至買方倉庫，則不須另投保；如承保範圍只至卸
貨港為止，則應安排自卸貨港口至買方倉庫之貨物運輸保險。

翻譯小試身手

❶ 若是您還沒建立線上帳戶，那就請您登記，然後我也會將您的帳戶設定為適用代理商折扣。

_____

_____

❷ 目前我們所有的信用卡客戶都已改為線上下單了。

_____

_____

參考答案   MP3 64

❶ If you do not have an online account set up yet, please register and then I will also set the Distributor discount to your account.

❷ All our credit card customers have switched to ordering online so far.

 E-mail 這樣寫 MP3 65

❶ Your Credit Memo is attached. You can use it as a deduction for future payments or use it to request a refund.

在此附上您的貸項通知單，您可以在下次付款時扣除此金額，或是用其來要求退款。

❷ To expedite exchanging for different product, we recommend returning for a refund and placing a new order. Please allow approximately 2 weeks for your refund to be processed.

為了要加速替換另一項產品給您，我們建議您辦理退貨、退款，另外再下一份新訂單給我們。我們約需兩個星期的時間來處理退款。

 英文書信這樣寫

Claim for Delay Shipment 裝運延誤索賠

Dear Tony,

This is to notify you that the delay of our P/O No. 3366 has disappointed us, especially the fact that we have not been given any explanation for the delay.

In accordance with our contract, the lots should have arrived at our port before July 7; however, we haven't received them yet. Due to the delay, we weren't able to meet the commitment to our customers. We shall ask you to be responsible for our losses caused by the delay.

Please arrange the delivery as soon as possible, as we won't accept the delay any more.

Sincerely yours,
Tom Smith

中文翻譯

湯尼 您好：

茲通知貴公司，我們公司第 3366 號訂單的延遲情形讓我們備感失望，尤其是貴公司尚未提供任何合理的解釋。

依據合約，此批貨應在七月七日前運抵我方港口，但截至目前，我們公司仍未收到貨物。因貴公司的延遲交貨，使我們公司違背了對客戶的承諾。此延誤所致的一切損失都將歸究於貴公司。

請貴公司儘速交貨，我們公司無法再接受任何延遲。

湯姆 史密斯 敬啟

國貿經驗談

▶▶ 貨櫃（Container）的「出口」作業流程：

- 訂艙：出口商依據合約向船公司或貨物代理辦理訂艙。
- 簽發裝箱單：確認訂艙後，由船公司簽發裝箱單，發送至貨櫃集散地安排空櫃和貨運交接。
- 發送空櫃：整櫃貨所需的空櫃，由船公司送交發貨人，拼櫃貨所需的空櫃通常由貨運站領取。
- 拼櫃貨裝櫃：貨櫃集散地依據訂艙單核收托運貨物，並簽發貨物收據後，在站內裝櫃。
- 整櫃貨裝櫃：發貨人收到空櫃後，自行裝櫃並準時送至貨櫃集散地。
- 貨物交接：站場收據為發貨人發貨和船公司收貨的憑證。
- 領取提單：發貨人憑站場收據，向船公司領取提單後向銀行結匯。
- 裝船：貨櫃集散地依據船舶積載計劃，進行裝船發運。

翻譯小試身手

❶ 在此將這次退款的交易識別碼列出如下，如有任何問題，再請告訴我。

_____

_____

❷ 若是您有要促銷文宣，您只需要點此連結，瀏覽我們所有的資料。（動物）吃（葉或嫩枝）

_____

_____

參考答案 MP3 66

❶ The transaction ID for this refund is listed below. Please let me know if you have any questions.

❷ If you need promotional flyers, you can simply click this link and browse through all our data.

34 付款延遲索賠 Claim for Delay Payment

 E-mail 這樣寫

❶ We would like to follow up with you about the progress we have made over the past month related to the integration of the CORE business into the ACE family. As you may know, CORE was acquired in March, 2016 by ACE, and we have been hard at work to ensure a smooth transition into our facility in Ohio.

要來跟您更新一下過去這幾個月來，我們整合柯爾公司到我們艾思集團的進度。可能您已經知道艾思在 2016 年三月已收購了柯爾公司，我們一直努力地做，為的就是要確保能夠順利將柯爾整合轉換到我們的俄亥俄州廠區。

❷ We are eager to combine ACE's resources with the already successful and innovative product line and staff at CORE.

我們現正積極地將艾思的資源與柯爾已經成功推行的創新產品線以及人員結合在一起。

英文書信這樣寫

Claim for Delay Payment 付款延遲索賠

Dear Tony,

　　We are very sorry for the late remittance of our payment. I'm writing to express our deep regret. The main reason was due to market sales not as expected We will transfer the amount to you by this weekend, and you can be sure that such a delay won't happen again.

　　Appreciate your understanding.

Sincerely yours,
Tom Smith

中文翻譯

湯尼 您好：

　　非常抱歉延遲匯付我們公司貨款。茲寫此文表達我們公司深摯歉意。主要是因市場銷售不如預期。我們會在本週末前匯款予貴司，並保證此匯款延遲不再發生。

　　感謝貴公司的理解。

湯姆 史密斯 敬啟

國貿經驗談

▶▶ 貨櫃（Container）的「進口」作業流程：

● 貨運單證：憑出口港寄來的有關貨運單證製作。

● 分發單證：將單證分別送交貨物代理、貨運站和貨櫃集散場。

● 到貨通知：通知收貨人有關船舶到港日，準備接貨，並於船舶到港後發出到貨通知。

● 提單：收貨人依據到貨通知，持正本提單向船公司或貨物代理換取提貨單。

● 提貨單：船公司或貨物代理核對正本提單無誤後，即簽發提貨單。

● 提貨：收貨人憑提貨單和進口許可證至貨櫃集散地辦理領櫃或提貨手續。

● 整櫃交貨：貨櫃集散地依據提貨單，將貨櫃交由收貨人。

● 拼櫃交貨：貨運站憑提貨單交貨。

翻譯小試身手

❶ 就是為了要讓您看看訂單程序有多麼簡單,我們製作了一段影片,可能對您有幫助。請點下列的圖示來看這段影片。

❷ 這份 Invoice 是供您付款參照用,若您有關於這筆訂單的任何問題,請向客戶服務人員提出詢問。

參考答案

 MP3 68

❶ Just to show you how easy the ordering process is, we have created a video which might be helpful. Please click the icon below to see the video.

❷ This Invoice is for your payment reference. Any questions regarding the order should be directed to customer service.

 E-mail 這樣寫 MP3 69

❶ If you do not accept the Terms and Conditions of Sale, please contact us to arrange for an immediate return of this un-opened product for a full refund.

若是您不接受我們的銷售條款，請與我們連絡，將您尚未開箱的產品立刻退回給我們，以辦理全額退款。

❷ The products should arrive on site ready for our in-house installation teams to get started before the delivery deadline. There will be a 1% penalty per day for delivery delay. The first 1% deduction starts immediately the day after the deadline.

貨品應在交貨期限前送抵我們的所在處，且應已準備就緒，可供我們內部安裝團隊開始作業。若有交貨延遲的情況發生，則將收取訂單金額的 1%為每日罰金，從期限後一天立刻開始起算。

英文書信這樣寫

Request of Replacement 要求換貨

Dear Tony,

　　Your delivery of our P/O No. 5678 arrived this morning. All the parts seem to be in good condition; however, there's a problem in assembling some of the parts with the connectors. After 100% assembly test, 52pcs out of 2,000pcs failed. Please arrange the replacement for the said quantities at once.

　　Your early reply would be appreciated.

Sincerely yours,
Tom Smith

中文翻譯

湯尼 您好：

　　貴公司寄送的第 5678 號訂單貨物已於今早抵達，所有貨物看來很良好，但是有些與連接管裝配時有問題。在進行 100%組裝測試後，發現這 2,000 件產品中有 52 件不合格。請立即安排更換產品。

　　若貴公司能早日回覆，我們公司將不勝感激。

湯姆 史密斯 敬啟

▶▶ 商品檢驗證書（Commodity Inspection Certificate）：

進出口商品經過商檢機構檢驗後，由該檢驗機構所出具的書面檢驗證明稱為「商品檢驗證書」。也可由生產單位，即製造商自行檢驗後出具檢驗報告，此類報告也可視為檢驗證書的一種。商品檢驗證書的作用主要有：

1. 為賣方所交付貨物的品質、重量、數量、包裝等是否符合合約規定的依據。
2. 當貨物存在爭議時，為買方對品質、數量、重量、包裝等提出拒收，並要求賠償的憑證。
3. 為記載貨物在裝卸時及運輸中的實際狀況，當貨物存在爭議時，為釐清責任歸屬的依據。
4. 為買賣雙方交接貨物、結算貨款和處理索賠的主要憑證。
5. 為繳付關稅、結算運費的憑證。
6. 為賣方向銀行議付貨款的單據。

在國際貿易中，檢驗機構可為國家設置的檢驗單位，或經由政府註冊的獨立檢驗公司，兩者的作用都是在對進出口的商品的裝運、質量、規格、包裝、數量、重量、衛生、安全、檢疫、殘損等進行檢驗和監督管理。進出口商品檢驗是貨物移轉過程中不可或缺的一個步驟。檢驗合格者，發給檢驗證書，出口商即可依此進行報關；檢驗不合格者，可申請複驗，複驗仍不合格者，則不得出口。

翻譯小試身手

❶ 這筆訂單預計在上方所列的出貨日當天或之前安排寄出。

_____

_____

❷ 在首頁裡的「產品標籤」已重新設計，使得在導覽上更有組織性可循。

_____

_____

參考答案

MP3 70

❶ The order is scheduled to be shipped on or before the date displayed above.

❷ The "products tab" located on the homepage has been given a makeover for more organized navigation.

Unit 36 付款延遲索賠（買賣索賠）Claim for Delay Payment with Seller

 E-mail 這樣寫

❶ The penalty amount is limited to a maximum of 10% of the amount of the order. In case of a delivery delay in excess of 4 weeks, we are entitled to declare the order null and void without any cost being charged for this.

罰款金額以訂單金額的 10%為上限，若交貨延遲的時間超過 4 個星期，那麼我們將有權宣稱此訂單無效，且也不能對我們收取任何費用。

❷ We develop world-class, cutting-edge chemicals for medical research use. Our mission is to accelerate research by providing the highest quality products, along with superior customer service and technical support.

我們研發世界級、最先進的化學製品供醫學研究使用，我們的使命是要提供最高品質的產品，以及優質的客戶服務與技術支援，以加快研究發展的速度。

英文書信這樣寫

Request of Return and Refund 要求退貨及退款

Dear Tony,

Please be informed that the goods against our P/O No. 3366 we received are not the type we ordered. I am afraid that you shipped the goods mistakenly.

I'll return the goods to you and ask for a refund.

Sincerely yours,
Tom Smith

中文翻譯

湯尼 您好：

茲通知貴公司，已收到的第 3366 號訂單貨物並非我們公司所訂購的型號。恐怕貴公司把貨物寄錯了。

我們公司將安排退貨並要求退款。

湯姆 史密斯 敬啟

▶▶ **資金短缺（shortage of finances）**是指企業所擁有的資金量少於維持企業正常營運所需要的資金量。資金為企業持續經營的必要條件，若資金短缺，又不能及時籌措，輕者將致使企業無法繼續商業活動，如採購生產原料，生產停擺、停工 等。重者將使企業償債能力下降，造成債務危機，進而影響企業信譽。

▶▶ **償債能力（debt-paying ability）**是指企業償還到期債務的承受能力或保證程度，包括償還短期債務和長期債務的能力，為企業能否持續生存和發展的關鍵。因此企業償債能力是評估企業財務狀況和經營能力的重要指標。

▶▶ 一般來說，買賣雙方在交易的初始時會在合約中定義付款期限，如買方因故無法在期限內履行支付款項的承諾，則賣方會追討應收帳款（及該收但仍未收回之款項），此商業行即為「逾期應收賬款管理」。雖然逾期應收賬款管理為相當普遍的商業行為，但為避免面臨帳款無法追收之風險，而成為呆帳，在買賣雙方開始首次交易前進行信用調查，即為相當重要的一環。

翻譯小試身手

❶ 請告訴我您要如何進行此訂單。

❷ 請見訂單單號 0925 的訂貨確認單如附，我們會依您要求，在今天出貨。

參考答案

MP3 72

❶ Please let me know how you would like to proceed with the order.

❷ Please see the attached order acknowledgment for order # 0925. The goods will be shipped today as requested.

37 Request of Order Cancellation 取消訂單

E-mail 這樣寫

MP3 73

❶ Our company was established in 1995. Since then we have offered high quality chemicals for research in various fields. We support our customers with a staff of highly experienced professionals as well as a modern, fully-equipped laboratory with capabilities to analyze the majority of the products we offer.

我們公司成立於 1995 年，一直以來都是提供高品質的化學製品，供各個不同領域的研究來使用，我們有極富經驗的專業人員來為我們的客戶提供服務，另外也擁有現代且配備完整的實驗室，以檢驗我們提供的多數產品。

❷ We hereby authorize HonTa Co., located at 10F, No. 10, Double-Tenth Rd., Sanchung Dist., New Taipei City, Taiwan, as the exclusive distributor for BASE products in Taiwan. BASE has provided the exclusive distribution right to HonTa Co. in Taiwan for distributing all BASE products. This letter expires on December 31, 2017 and authorizes HonTa Co. to quote, enter into contracts, government tenders and supply the products on our behalf.

我們在此授權位於新北市三重區雙十路十號十樓的宏大公司,為貝斯產品的臺灣獨家代理商。貝斯提供臺灣的獨家代理權給宏大公司,讓其經銷所有貝斯的產品。此授權書的效期至 2017 年十二月三十一日為止,授權宏大代表我們提供報價、簽立合約、參與政府標案,以及供應產品。

 英文書信這樣寫

Request of Order Cancellation 取消訂單

Dear Tony,

I have to notify you that we are cancelling our P/O No. 7788, as we can't accept such a long postponement.

The delay of the goods has caused great loss to our business. In accordance with the terms of our contract, we are obtaining the rights to ask for compensation due to the delay more than two weeks. Shall there be any questions, please let me know.

Sincerely yours,
Tom Smith

湯尼 您好：

　　必須通知貴公司，我們公司將取消第 7788 號訂單事宜。我們無法接受長時間延遲交貨。

　　貨物的延遲已對我們公司業務造成極大損失。依據雙方合約條款，延誤超過兩週以上，我們公司有權請求賠償。如有任何疑問，請讓找知悉。

湯姆 史密斯 敬啟

國貿經驗談

▶▶ **商品檢驗證書（Commodity Inspection Certificate）：**

進出口商品經過商檢機構檢驗後，由該檢驗機構所出具的書面檢驗證明稱為「商品檢驗證書」。也可由生產單位，即製造商自行檢驗後出具檢驗報告，此類報告也可視為檢驗證書的一種。商品檢驗證書的作用主要有：

- 為賣方所交付貨物的品質、重量、數量、包裝等是否符合合約規定的依據。
- 當貨物存在爭議時，為買方對品質、數量、重量、包裝等提出拒收，並要求賠償的憑證。
- 為記載貨物在裝卸時及運輸中的實際狀況，當貨物存在爭議時，為釐清責任歸屬的依據。
- 為買賣雙方交接貨物、結算貨款和處理索賠的主要憑證。

● 為繳付關稅、結算運費的憑證。

● 為賣方向銀行議付貨款的單據。

翻譯小試身手

❶ 所有國際訂單皆須以 e-mail 回覆確認正確無誤後，我們才會出貨。

_____

_____

❷ 我們只接受貨物有瑕疵或與訂單不符的退貨。

_____

_____

參考答案  MP3 74

❶ All International Orders must be confirmed as accurate before shipping by sending an email to us.

❷ We only accept returns where goods are faulty or where they do not conform to *the order.*

Unit

38 要求換貨 Request of Replacement

 E-mail 這樣寫 MP3 75

❶ Our policy is not to give exclusivity. However, we could consider an exclusive arrangement if a minimum order value was agreed upon, e.g. US$ 200,000, whereby any difference in amount ordered and the agreed upon target would be invoiced to you at the end of the year.

我們公司的政策是不給獨家代理權，不過，若是您同意最低訂單金額的要求，例如 US$ 200,000，而且接受所訂金額與所同意目標的差額可在年底開發票給您，那我們就可考慮獨家代理的安排。

❷ For now we don't have any intentions to provide a sole distributorship to any of the Taiwanese distributors. I appreciate the progress that you exhibit with our sales, but the current situation is that none of the distributors exhibit a significant superiority over the other.

目前我們並沒有意願給任何一家台灣代理商獨家代理權，我很感謝您在銷售我們產品上所展現的成長，但現在的狀況是沒有一家代理商在表現上有明顯優於另一家代理商。

英文書信這樣寫

Extend the Time Limit of Claim 延期索賠

Dear Tom,

Please accept our sincere apologies for not replying to your claim request promptly, as I have been preparing for the quality meeting to discuss the functionality problem raised by you.

Therefore, please grant us more days to work out the solution and extend the time limit of claim to the end of this month.

Please feel free to let me know if you have any concerns.

Sincerely yours,
Tony Yang

中文翻譯

湯姆 您好：

　　未能及時回覆貴公司的索賠請求，請接受我們公司誠摯的歉意。乃因本人持續忙於安排品質會議討論貴公司提出的功能異常問題。

因此，請特准我們公司幾日時間找出解決方案，並延長索賠期限至本月底。

如貴公司有任何考量，請不吝告知。

湯尼 揚 敬啟

國貿經驗談

▶▶ 在國際貿易中，檢驗機構可為國家設置的檢驗單位，或經由政府註冊的獨立檢驗公司，兩者的作用都是在對進出口的商品的裝運、質量、規格、包裝、數量、重量、衛生、安全、檢疫、殘損等進行檢驗和監督管理。進出口商品檢驗是貨物移轉過程中不可或缺的一個步驟。檢驗合格者，發給檢驗證書，出口商即可依此進行報關；檢驗不合格者，可申請複驗，複驗仍不合格者，則不得出口。

翻譯小試身手

❶ 請檢查附件的訂貨確認單明細，若有任何不符之處，還請立刻告訴我。

❷ 請注意，若我們沒有收到 e-mail 回覆確認，所有的訂單都會扣留著不處理。

參考答案 MP3 76

❶ Please kindly check the OC details as attached and let me know immediately if there is any discrepancy.

❷ Please note that without receiving confirmations by return e-mail, all orders will be held and not processed.

2 Part 國貿進階篇

國貿進階篇規劃了「國貿經驗談」、「關鍵句型」、「英文書信這樣寫」和「翻譯小試身手」。可以先由每單元的國貿經驗談下手，讀讀中文，接著學習兩組關鍵句型，練習基礎造句，逐步進階到短篇英文書信撰寫。仿效書信中文句，也可藉由書籍CD的便利貼光碟輕鬆複製相關文句，省下撰寫書信的時間。最後，藉由翻譯小試身手練習四句相關國貿用句，在國貿相關證照跟中英翻譯考試上都能獲取佳績。

Unit 01 新品上市 Notification of New Product Launch

國貿經驗談

▶▶ 隨著時代變遷、科技進步及網際網路的普及，現行商業活動已鮮少使用郵寄信件作為溝通往來方式，大多藉由電子郵件作為溝通工具，甚至隨著智慧型手機的廣泛運用，有些更利用 Line, WhatsApp, WeChat 等軟體達到即時溝通的目的。無論使用何種溝通工具，務必切記 "口說無憑"，在雙方進行討論後，建議補上一封電子郵寄詳載雙方討論過之內容，作為依據。

🔍 關鍵句型

be successful in sth 在某方面獲得成功

（例句說明）

· Jack is highly successful in business.

傑克經商相當成功.

· Our company was successful in negotiating lower prices.

我們公司很成功地將價格協商成較低價格。

（替換句型）

· Our company achieved success in negotiating lower prices.

Sb. be much more concerned about sth 某人更在意某事

（例句說明）

· Customers are much more concerned about product quality.

客戶更在意產品質量。

· My supervisor is much more concerned about my performance.

主管更在意我的表現。

（替換句型）

· My supervisor is much more conscious about my performance.

Dear Sir and Madam,

We're very pleased to inform you about our new balancing valve designed by our R&D team.

The new type has improved the function to be more stable and efficient. In view of high-quality and low-price, we think that the new type will be very competitive and well received in the market.

Please never hesitate to contact us if you'd like to know more about production information.

Yours sincerely,
Wesley Yang

中文翻譯

敬啟者：

我們公司非常高興通知您，我們研發部門開發出了新型平衡閥。

新型產品改良了產品功能，使其更穩定及更有效率。基於物美價廉，相信它將很有競爭力，且在市場上受到青睞。

如欲瞭解更多產品資訊，請儘管與我們公司聯絡。

衛斯理 楊 敬啟

❶ 因為技術上遇到困難,我們今天沒辦法出貨,到貨日也會往後延。

_____

_____

❷ 請明確地告知您所要的是哪一種包裝,這樣我才能修改報價單給您。

_____

_____

❸ 此交易驗證失敗,因此交易已中止,請以別張信用卡再試一次。

_____

_____

❹ 這份合約會在收到您簽回訂單或是書面表示接受後生效。

_____

_____

參考答案   MP3 77

❶ Due to technical difficulties, we will not be able to ship your package today and will postpone the delivery date.

❷ Please inform specifically what packaging you would like and I can supply a revised quote.

❸ Authentication of this transaction failed and the transaction has been suspended. Please try again using a different credit card.

❹ The Contract shall come into effect upon receipt of a signed order form or written acceptance.

02 產品停產 Notification of Production Suspense

國貿經驗談

▶▶ 「產品停產通知」為商業書信中重要的書信之一,產品停產將可能引起客戶端供貨短缺,也關係著供應商與客戶間是否會持續合作關係。因此產品停產通知除了要詳載停產的產品品號、顏色、規格及停產日期,並告知停產的原因及最後下單日期,讓客戶能及早鋪貨及備庫。另外,最好能提供 "補救方案",例如推薦類似規格的既有產品或預告即將開發的等同規格新品等,以維持與客戶後續之合作關係。

關鍵句型

be difficult for us to accept 難以接受

例句說明

· Your offer is too high, which is difficult for us to accept.

貴司的報價太高了, 我方難以接受。

· I'm afraid it will be difficult for us to reduce the prices.

我恐怕降價會有所困難。

替換句型

· I can't adapt well to your high offer.

Sb. be delighted to 樂於

例句說明

· I'd be absolutely delighted to visit your company.

我非常樂意去拜訪貴司。

· Kate was delighted to meet you.

凱特非常高興見到您。

替換句型

· Kate was glad to meet you.

Dear Customers,

This notification is about the end of production.

Due to the shortage of raw materials, we have decided to discontinue three models of ball valve, said P/N 0011, 0012, and 0013. Taking into account the demands of the market, we expect to launch out new product in this series in the second quarter. We'll keep you advised of the related information, once available.

Best Industrial is deeply grateful for your support of these products.

Yours Sincerely,
Wesley Yang

中文翻譯

親愛的客戶，

　　特此奉告產品停產。

　　由於原物料短缺，我司決定停產三款球閥產品，分別是產品編號 0011, 0012, 以及 0013。考量到市場需求，我司預計在第二季上市同系列新品。待新品完成，我司將會通知您相關訊息。

　　貝斯特工業深摯地感激您對該產品的支持。

衛斯理 楊 敬啟

翻譯小試身手

❶ 開始生產前，須預先付清貨款，等我們一收到貨款，就會立即開始處理訂單。

_____

_____

❷ 對於訂製合成的訂單，您得先行支付 30％的貨款，之後我們才會開始處理訂單。

_____

_____

❸ 請務必在訂製規格表上列明您客戶所要的產品與確實的數量。

_____

_____

❹ 您所詢的產品並沒有現貨，我們預期六月可有新批推出，請告知您的客戶是否願意等。

_____

_____

參考答案　　　　　　　　　　　　　　　　MP3 78

❶ Advance payment is required prior to the commencement of production. Once the payment is received, we'll process your order right away.

❷ For custom synthesis orders, a 30% advance payment is needed to initiate the order.

❸ Please make sure that you indicate the products and exact quantity that your customer will require on the custom specification form.

❹ Your inquired product is out of stock. We expect a new release in June. Please let me know whether your customer is willing to wait or not.

Unit

03 恢復往來
Business Resuming

國貿經驗談

▶▶ 對於許久無業務往來之客戶以問候對方為優先，以不至於令對方
覺得唐突，而後表達關切中斷往來之原因，例如是否我方服務不
周或者對方業務項目變更等。最後進而表達期盼持續合作關係之
渴望。

 **關鍵句型**

what have you been up to
在忙些什麼?(表達問候對方近來如何)

例句說明

· What have you been up to recently?

最近在忙些什麼?

替換句型

· What have you been doing recently?

· What has kept you busy recently?

· What are you occupied recently?

give priority to 優先考慮

例句說明

· Our next goal should give priority to the development of new product.

我們下一步要優先發展新品。

· To avoid quality defect, we should give priority to root cause.

避免品劣品需先找出根源。

替換句型

· The development of new product is a top priority as our next goal.

Dear Amanda,

How is everything? I re-ran in my mind our accounts for the past year, and note that we have not received neither information nor orders from you for a long time.

Assuming that your company is still operating in the series of our production, please inform us of your recent sales intention and goal of our reference. In case you have any comments or suggestions about our products to order, please do not hesitate to put forward to our company, for the benefit of our company as review and improvement.

We presume that your company must be willing to learn that our products make a series of upgrade, both in technology and functions. Enclosed are the updated catalogue and price list, including the new models of the line you used to order. You'll find out our products are in line with your needs just well and be quite satisfied with them. Furthermore, the same series of samples have been in transit for your inspection.

We look forward to receiving your positive response and resuming our friendly business connections.

Yours sincerely,
Wesley Yang

中文翻譯

艾曼達 您好，

　　一切都好嗎？我在腦海中回顧雙方過往的紀錄，發現我司已有很長一段時間沒有收到貴公司的訊息及訂單了。

　　假設貴公司仍在經營我司生產的系列產品，請告知貴公司近期的銷售目的和目標，供我們公司參考。若貴公司對向我司訂購產品有任何意見及建議，請不吝向我們公司提出，以利我司作為檢討及改善的參考

　　相信貴公司一定樂於獲知我司的產品在技術上及功能上做出了一系列的提升。附件是更新的產品目錄及價格表，其中包含貴公司以往訂購的系列產品的新款式。貴公司將會發現我司產品正符合貴司需要，並會對他們相當滿意。此外同一系列樣品已寄出予貴公司檢視。

　　我司期待收到貴公司回復，以及雙方恢復友好業務關係。

衛斯理 楊 敬啟

❶ 我們有信心可以確實生產符合您規格要求的產品。

❷ 在即將舉辦的會議中，我們打算在我們的攤位上協調安排個短短的代理商會面時間。

❸ 我們所有產品都是走 FedEx 國際優先型的方式出貨，除非客戶有自己配合的快遞公司。

❹ 若是您要我向聯邦快遞要求報來國際經濟型空運或海運的運費，就請告訴我。

參考答案   MP3 79

❶ We are confident that we can actually make the product that meets your specification requirements.

❷ We plan to coordinate a short distributor meetup at our booth for the coming conference.

❸ We ship all products through FedEx International Priority unless the customer has their own courier.

❹ If you need me to solicit a FedEx International Economy Air or Sea Freight quote for you, please just let me know.

Unit 04 同意代理 Agreement for Exclusive Agency

 國貿經驗談

▶▶ 代理商的種類概述：

▶▶ 獨家代理商（sole agent）：指製造商在同一區域內只能授權單一代理商，但製造商本身仍可在此區域進行銷售行為。

▶▶ 非獨家代理（simple agent）： 同一區域內可同時有多家代理商，製造商本身亦可在此區域進行銷售行為。

▶▶ 絕對排他代理商（exclusive agent）： 指製造商在同一區域內只能授權單一代理商，且製造商本身不可在此區域進行銷售行為。

🔍 關鍵句型

be aware of sth 知道某事

例句說明

· Our customers are well aware of our prices.

客戶十分瞭解我們的價格。

· Be aware of the registration deadline.

注意報名截止日。

替換句型

· Be conscious of the registration deadline.

to one's advantage 對某人有利

例句說明

· High quality will be to your advantage in competition.

高品質使你有競爭優勢。

· The contract works to buyer's best advantage.

該合約對買方最得利。

替換句型

· You benefit greatly from high quality.

Dear Sirs and Madam,

The object of this letter is to express that we are eager for being your sole agent in EU market.

World Enterprises is a reliable company with rich experience in the line of the industrial part with over a hundred years. Having had a professional team of sales representatives and excellent show rooms, we hereby recommend ourselves to act as your sole agent for your balancing valve in EU market.

Enclosed is our proposal agency agreement for your reference. Please feel free to contact us for further discussion of the clauses.

Thank you for your time on reading this email. We are looking forward to your favorable reply by return.

Yours sincerely,
Scott Fuller

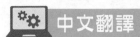
中文翻譯

敬啟者：

　　特此表達我們公司熱切希望成為貴公司在歐盟市產的獨家代理商。

　　世界企業為一家信譽卓越的公司，在此行業擁有逾百年的豐富經驗。我們擁有專業的銷售團隊及一流的陳列室，在此自我推薦作為貴公司平衡閥在歐盟市場的獨家代理商。

　　請參閱附件我們公司提案的代理權合約。請隨時與我們聯繫，進一步討論合約條款。

　　感謝您撥冗審閱此文。靜候佳音！

史考特 富勒 敬啟

❶ 這筆訂單將會在下星期三前,以聯邦快遞 P1 國際服務型的方式出貨。

❷ 若是你選擇透過貨物承攬業務代理公司(簡稱貨代)辦理出貨,請注意您就必須負責安排所有的文書作業與取貨事宜。

❸ 請放心,品質一向是我們的首要堅持,我們會確定產品品質在運送途中不會受到影響。

❹ 儲存時請注意要將容器緊緊關好。

參考答案

MP3 80

❶ The order will be shipped via Federal Express Priority First International Service by next Wednesday.

❷ If you choose a freight forwarder option, please note that you will be responsible for coordinating all paperwork and scheduling the pick-up.

❸ Please rest assured that the quality is always our first priority. We'll make sure that the product quality will not be affected during transport.

❹ Please note that you must keep the container tightly closed when storing.

Unit 05
產品介紹（對首次見面客戶）Product Introduction to New Customer

 國貿經驗談

▶▶ 對於未合作過的新客戶提出需求時，不同於對既有客戶著重在詳述特定產品規格，可藉此介紹公司服務項目及優勢，強調生產技術、品質系統及價格條件等吸引客戶，開啟首次合作的契機。

關鍵句型

I'd like to introduce 我想介紹

例句說明

· I'd like to introduce you to our CEO.

我想介紹你認識我們的執行長。

· I'd like to introduce myself to you.

我想向你作自我介紹。

替換句型

· Allow me to present my CEO to you.

Sb. successfully challenge for 成功挑戰

例句說明

· I successfully challenged for the project award.

我成功爭取到此專案。

· He successfully challenged for the chief salesman.

他成功成為首席業務。

替換句型

· I succeed in getting the project award.

Dear Mr. Nate,

We are very pleased to receive your call, knowing that you are interested in our products.

Let us introduce ourselves as a manufacturer and exporter, establishing a good reputation in the field of industrial valves at home and abroad. We can provide customers around the world with high quality service and reliable quality, covering manufacturing, quality inspection, foreign trade, warehousing, and logistics field.

Over the years, our company set up a complete set of product system, containing a variety of heating valves, check valve, BS standard valve, cut-off valve. Please see the attachment for the latest catalogue and price list. We believe that our professional technology and quality inspection system will bring us a win-win. In addition, it is worth mentioning that, we can provide you with a competitive price.

Looking forward to receiving a favorable reply soon.

Thanks & best regards,

Wesley Yang
Product Manager
Best International Trade Corp.
Tel: 886-2-22335555 / Mobile: 886-958-223355

 中文翻譯

奈特先生 您好，

　　很高興接到您的來電，得知貴公司對我們的產品感興趣。

　　請容在下介紹我們公司，為在國內和國外工業閥門領域建立了良好的信譽製造商及出口商。我們可以提供世界各地的客戶，高品質的服務和可靠的品質，涵蓋生產製造，品質檢驗，對外貿易，倉儲物流等領域。

　　多年來，我們公司已擁有一套完整的產品體系，包含各種供暖閥門，檢查底閥，BS 標準閥，截止閥等。請參閱附件最新的產品目錄及報價表。相信我們公司的專業技能及品檢系統會為我們帶來雙贏。此外，值得一提的是，我們可以為您提供具有競爭力的價格。

　　靜候佳音!
　　衛斯理 楊 敬啟
　　產品經理
　　倍斯特國際貿易公司
　　電話：886-2-22335555
　　手機：886-958-223355

翻譯小試身手

❶ 我們還不知道確切的運費是多少，因為費率要用尺寸大小和重量
來算。

_____

_____

❷ 這箱子的毛重估算起來差不多有 50～60 磅。

_____

_____

❸ 合適的包裝有助於確保您的貨品可安全抵達。

_____

_____

❹ 這項產品對光線敏感，因此應避免直接日曬及紫外線光源。

_____

_____

參考答案   MP3 81

❶ We don't know exactly how much the freight is because the rate is calculated based on dimension and weight.

❷ The roughly estimated gross weight of this box is 50~60 lbs.

❸ Proper packaging can help ensure that your goods arrive safely.

❹ The product material is light sensitive and should therefore be protected from direct sunlight and UV sources.

 國貿經驗談

▶▶ 對於未合作過的新客戶提出需求時，不同於對既有客戶著重在詳述特定產品規格，可藉此介紹公司服務項目及優勢，強調生產技術、品質系統及價格條件等吸引客戶，開啟首次合作的契機。

🔍 關鍵句型

I'd like to introduce 我想介紹

例句說明

· I'd like to introduce you to our CEO.

我想介紹你認識我們的執行長。

· I'd like to introduce myself to you.

我想向你作自我介紹。

替換句型

· Allow me to present my CEO to you.

Sb. successfully challenge for 成功挑戰

例句說明

· I successfully challenged for the project award.

我成功爭取到此專案。

· He successfully challenged for the chief salesman.

他成功成為首席業務。

替換句型

· I succeed in getting the project award.

英文書信這樣寫

Dear Ms. Ryder,

Good morning! It is our highly pleasure to learn your honored name through ABC Co, and know that you are looking for industrial valve.

Please allow me to take this opportunity to introduce ourselves - Best Group. - TOP manufacturer and exporter of industrial valves with the leading manufacturing capabilities in Taiwan. Our business involves manufacture, quality inspection, foreign trade, storage, and logistics. We can provide high-quality service and credible quality to the clients all over the world. Our products mainly focus on industrial valves, such as all kinds of heating valves, check & foot valves, BS standard valves, globe valves and so on. Especially we can provide customized R&D services, according to your drawings /samples to meet your special requirements. Please visit our website www.best.com.tw to know details of our company and the full range of our product.

Thanks for your valuable time to read this letter, and we are looking forward to establishing a long-term & mutual beneficial partnership with you in the near future.

Thanks & best regards,
Wesley Yang
Product Manager
Best International Trade Corp.
Tel: 886-2-22335555 / Mobile: 886-958-223355

中文翻譯

瑞德 小姐 您好，

日安！我們公司非常榮幸透過 ABC 公司的介紹得知貴公司的尊名，並知道貴公司正在尋找工業閥門。

請容許我借此機會介紹我們公司 – 倍斯特集團 - 台灣具備領先工業閥門製造能力的頂尖製造商及出口廠商。我們的業務涉及製造、品質檢驗、對外貿易，以及倉儲物流。我們可以提供世界各地的客戶高品質的服務及可靠質量的產品。產品主要集中於工業閥門：如各種供暖閥門，檢查底閥，BS 標準閥，截止閥等。特別是我們可以依據您的圖紙或樣品，提供客製化的研發服務來滿足您的特殊要求。請瀏覽找司的網站 www.best.com.tw，以瞭解詳細公司資訊和全面的產品細節。

感謝您撥冗閱讀此封郵件，我們公司期待在不久的將來與貴公司建立長期互惠互利的合作夥伴關係。

衛斯理 楊 敬啟
產品經理

Part 1 國貿基礎篇

Part 2 國貿進階篇

倍斯特國際貿易公司

電話：886-2-22335555

手機：886-958-223355

翻譯小試身手

❶ 這項產品對溫度敏感，因此，請確認包裹有儲存在合適的溫度之下。

❷ 我們在商業發票上都是列出產品的真實金額，以辦理清關。

❸ 請告訴我除了商業發票與裝箱單之外，您還需要哪些其他的文件（如提單、出貨人的指示信函等）。

❹ 空運主提單是由航空運輸公司出具，而不是貨代。

參考答案 MP3 82

❶ This product is temperature sensitive and; therefore, please make sure that the parcel is being stored at the appropriate temperature.

❷ We always declare the true value of our products on Commercial Invoices for Customs clearance.

❸ Please let me know what additional paperwork aside from the commercial invoice and packing list (i.e. bill of lading, Shipper's Letter of Instruction, etc.) you require.

❹ A Master Air Waybill (MAWB) is issued by the carrier and not the freight forwarder.

Unit

07 信用調查
Credit Investigation

 國貿經驗談

▶▶ 企業間在進行交易前進行徵信極為重要,尤其是針對新客戶、久未往來之客戶及位於政治環境動盪國家之客戶等。信用調查的內容主要針對財務條件、信用狀況及經營能力等方面進行調查,調查的目的無非就是為了避免買賣雙方交易後無法獲得應有利益。

▶▶ 選擇的信用調查機構如果是金融業者,最好是行庫對行庫,例如我方往來銀行向對方往來銀行提出,不論是透過哪一個信用調查機構,都須提供給調查機構有關授信者的完整公司名稱及住址,以及其往來行庫名稱與帳號等。我方在正式啟動交易前,可向對方要求提供上述資料,而一般信譽良好的企業甚至會主動提供相關資料。

關鍵句型

fall into arrears with 拖欠

例句說明

· The pledged realty has been expropriated, as your company fell into arrears with the loan

因貴司拖欠貸款，貴司抵押之不動產已被沒收。

· Your company have fallen into arrears with three-month's rent.

貴司已拖欠三個月租金。

替換句型

· Your company have fallen behind with three-month's rent.

be assured that 放心

例句說明

· You can be assured that the shipment passed 100% inspection.

你放心這筆貨通過百分百全檢合格。

· How can we be assured that the shipment have been fully inspected?

我們怎麼能確信該批貨經過完整檢驗?

替換句型

· You can set your mind at rest that the shipment passed 100% inspection.

Dear Sir,

Great Inc. with details under-mentioned recently approached to have business dealings with us and referred us to your bank.

We shall appreciate your bank to provide us related information of their financial and business status, and meet all the expenses incurred in this connection upon being notified.

Please accept our apology for any trouble and inconvenience we may cause you. You may rest assured that any information provided will be conducted entirely in confidential, without responsibility on your party.

Yours sincerely,
Wesley Yang

 中文翻譯

敬啟者：

　　偉大股份有限公司近期與我們公司接洽業務往來事宜，並向我們公司推薦貴行進行諮詢。

　　我們公司感謝貴行能提供有關偉大公司的財務和業務狀況的相關資訊，在被告知之情形下，我們公司將負擔所有因此產生之費用。

　　對於可能帶來貴行任何麻煩和不便，請接受我們公司的歉意。請放心，貴行提供的任何資訊將完全保密，你方不負任何責任。

衛斯理 楊 敬啟

❶ 我們今天將會以 FedEx 快遞寄送正本聲明函給你,不會另外收費,因為是我們主管機關的疏忽,所以他們同意免費出具。

❷ 提醒您一下,請將裝箱單正本附在所出的貨裡。

❸ 這張提單是您快遞貨物的票證與護照,以確保能及時、準確並安全地出貨。

❹ 出貨人可申報較高的運輸價值並支付附加費用,以擴充責任限度。

參考答案   MP3 83

❶ We will send the original statement to you today via Federal Express. There will not be an additional charge as our authority agreed to waive the fee due to their overlook of the statement.

❷ Please be reminded to enclose the original Packing Slip with the shipment.

❸ The waybill is your express shipment's ticket and passport to ensure timely, accurate and secure delivery.

❹ Shipper may increase the limitation of liability by declaring a higher value for carriage and paying a supplemental charge if required.

Unit

08 產品評估
Product Evaluation

 國貿經驗談

在前一章節中説明了書信的基本架構,在這裡我們將詳細説明商業書信中常用的稱謂,即「收件人稱呼(Salutation)」。

● 統稱:適用於統一信函寄發給多位客戶,如尾牙邀請函、休假公告等。例如:Dear Sir(親愛的先生)、Dear Madam(親愛的女士)、Dear Sir and Madam(敬啟者)、Dear Customers(親愛的客戶)。

● 姓氏 + 先生/女士之稱謂:適用於較不熟悉或首次聯繫之客戶。例如 Dear Mr. Lee(親愛的李先生)、Dear Miss/Ms. Lee(親愛的李小姐/女士)、Dear Mrs. Lee(親愛的李夫人)、Dear Mr. and Mrs.(親愛的李氏賢伉儷)。

● 姓名 + 先生/女士之稱謂:適用於較熟悉之客戶。例如 Dear Mr. John Lee(親愛的李約翰先生)、Dear Miss/Ms. Annie Lee(親愛的李安妮小姐)。

● 名 不加任何稱謂:適用於長期合作、關係密切之客戶。例如 Dear John(親愛的約翰)、Dear Annie(親愛的安妮)。

 關鍵句型

Sb. need to pay particular attention to sth
某人需特別關注某事

例句說明

· The engineer needs to pay particular attention to the tooling building status to ensure submission on time.
工程師需特別注意模具建構情形，以確保準時送樣。

· Buyers needs to pay particular attention to the material cost, so that they can request cost reduction from sellers accordingly.
買方需特別注意原物料價格，才能依此要求賣方降價。

替換句型

· Obviously, the material costs are the key point that the buyers will focus on when raise up the proposal of cost reduction to sellers.

There is barely time to V 來不及做某事

例句說明

· There is barely time to request sellers to revise the piece price based on the updated material cost.
來不及要求賣方依據最新原物料價格修改產品單價。

· There is barely time to implement 100% inspection before release.

沒有足夠的時間在出貨前進行全檢。

· It's too late to implement 100% inspection before release.

 英文書信這樣寫

Dear Olivia,

We have duly received your inquiry of your part number A5197 by email dated Jan. 10th, and took the liberty to enclose the quotation sheet herein. Meanwhile, you may refer to our proposal engineering suggestions as the drawings attached.

If there's no further comment, please revise our engineering drawing proposal by return and release the tooling as well as sample order to kick off tooling building and sample preparation. Or else, please share your concern for our reference.

We hope to receive your reply without any delay to kick off new product development immediately to meet your urgent demand.

Yours sincerely,
Wesley Yang

中文翻譯

奧莉維亞 您好：

我們已正式透過貴公司一月十日的電郵收到貴公司編號 A5197 的詢價，在此提供附件報價單。同時，貴司可以參考我司建議的工程建議如附圖。

如果貴公司無進一步的評論，請依據我們公司工程建議改回圖紙並下模具及樣品訂單，以啟動模具製造和樣品製備。否則，請提供您的考量點供我們公司參考。

我們公司期盼收到您立即的回覆，以即刻啟動新產品開發來因應貴公司急迫的需求。

衛斯理楊 敬啟

翻譯小試身手

1 能源商品的市場極為複雜多變，而且會受地緣政治因素所影響。

_____

_____

2 這則聲明應要在進口證上列出，若沒有的話，則收貨人（進口人）就必須另行出具一份證明。

_____

_____

3 請注意，您必須全額負擔清關所收取的任何費用。

_____

_____

4 提單中所申報的貨品狀況良好，可供運送。

_____

_____

參考答案　MP3 84

❶ The energy commodity market is extremely volatile and driven by geopolitical factors.

❷ This statement may be included on the import certificate; however, if it is not, the consignee (importer) must make a certification.

❸ Please note that you are solely responsible for any customs clearance fees required.

❹ The goods declared on the Air Waybill are apparently in good order and condition for carriage.

Unit 09

產品開發
Product Development

國貿經驗談

▶▶ PPAP（Production Part Approval Process）生產件批准程式：
PPAP 的目的是用來確定供應商是否已經正確理解了顧客工程設計和規範的要求，以及供應商生產過程是否具有潛在能力生產該產品，並在實際生產過程中按所規定的生產步驟來滿足顧客要求的產品質量。

以上資料來源參考智庫百科網頁 http://www.mbalib.com

關鍵句型

A take the liberty of V-ing B
A 藉此機會（冒昧地/自作主張地）對 B

例句說明

· May I take the liberty of asking you to provide some samples?
我可以冒昧地 您提供一些樣品嗎?

· The supplier took the liberty of sending me some samples.
供應商主動寄給我一些樣品。

替換句型

· The supplier took the advantage of this opportunity to send me some samples.

In accordance with the agreement contained in one's favor of (date) 依據某人某日來函同意

例句說明

· In accordance with the agreement contained in the buyer's favor of May 1ˢᵗ, the air shipment will be effected with freight collect.
依據買方五月一日來函同意，該批空運貨物將以運費到付出貨。

替換句型

· According to the agreement given in David's mail dated May 1ˢᵗ, the part cost will be reduced by 3% from next month.

Dear Wesley,

<u>We take the liberty of informing you that</u> part number A5197 is awarded to Best Co. based on your quotation sheet of Jan. 11 and the engineering proposal was mutually.

Enclosed please find the final 2D & 3D file, tooling order, and sample order. <u>In accordance with the agreement contained in your favor of Jan. 12</u>, please follow the timeframe of new product development as below.
- Estimated date of tooling completion: Feb. 2.
- Estimated date of sample completion: Feb. 12

In addition, a kick-off meeting is suggested to be held for review and discussion of proposed engineering issues to smooth the tooling building as well as sampling. We'll try to arrange a mutually convenient time and give you a notification soon.

Any question, please let me know.

Yours sincerely,
Olivia Porter

 中文翻譯

衛斯理 您好：

　　我們公司通知您，產品編號 A5197 將依據貴公司一月十一日報價表及雙方同意之工程方案，委由貴公司開發生產。

　　隨函附上最終的 2D 和 3D 檔案、模具訂單，以及樣品訂單。依據貴公司一月十二日所同意，請遵循新產品開發時程如下：

　　預估模具完成日：二月二日。
　　預估樣品完成日：二月十二日

　　此外，建議舉行產品啟始會議檢視及討論提出的工程議題，以使開模及打樣順利進行。我們會儘可能安排一個雙方都方便的時間，並很快地給您一個通知。

　　如果有任何問題，請讓找知道。

奧莉維亞波特 敬啟

❶ ABA 路徑號碼是美國銀行協會設定給金融機構的九碼辨識號碼。

❷ 這些產品會在塑袋內包裝上有張 GHS（化學品分類與標示之全球調和系統）標籤。

❸ 我們的產品不需要 DEFRA 證明來清關，但請跟我們確認是否海關有需要任何其他的聲明。

❹ 若是此次出貨您可提供您們自己的 FedEx 帳號，那我們就不會跟您收運費和稅負費用。

❶ The *ABA* routing number is a 9-digit identification number assigned to financial institutions by the American Bankers Association (ABA).

❷ These products will have a GHS (Globally Harmonized System of Classification and Labeling of Chemicals) sticker on the inner packaging of the plastic bag.

❸ Our products do not require a DEFRA1 Certificate for customs purposes, but please confirm if they require any other kind of declaration.

❹ We will waive both the freight and duty / tax charges if you can provide your own FedEx account for the shipment.

Unit

10 送樣審核
PPAP Verification

 國貿經驗談

▶▶ 簡而言之，所謂 PPAP（Production Part Approval Process）生產件批准程式，就是產品在正式量產前，供應商或製造商為了證明自家的生產製程有能力生產該產品，且生產的產品能符合客戶不論是製程、尺寸、外觀、功能等等規範及要求，而將打樣樣品遞交給客戶端審核之流程。送樣審核之要求依各客戶定義之，但一般來說送樣核可後（PPAP Approval），客戶會發予核可證書（PSW, Part Submission Warrant），作為供應商或製造商後續啟動大量生產的依據；反之，則送樣判退（PPAP Rejection），須依據客戶要求重新送樣（Re-PPAP Submission）。

🔍 關鍵句型

A very much regret to inform B　A 十分抱歉通知 B

例句說明

- I very much regret to inform you that we are unable to accept your kind invitation of the end-year party.

  我十分抱歉通知您我們公司不克接受貴公司年終尾牙派對的盛情邀約。

- My boss very much regretted to inform our team of layoffs notice.

  老闆十分遺憾的通知組員有關裁員的通知。

替換句型

- It is with the greatest regret that my boss informed our team of layoffs notice.

In response to mail/calling of date 回覆某日來函/來電

例句說明

- In response to your mail yesterday, please arrange the air shipment with freight prepaid.

  茲回覆您昨日來函,請安排貴公司付費空運。

- In response to Erica's mail of March 23, we can only accept the air shipment with freight collect.

  茲回覆艾瑞卡三月二十三日來函,我們公司僅能接受運費到付空運。

· Answering Erica's mail of March 23, we can only accept the air shipment with freight collection.

英文書信這樣寫

Dear Wesley,

    <u>In response to</u> your requirement by yesterday's call, <u>I very much regret to inform you that</u> our engineering team can't accept the critical dimension adjusted to 2.552 for part number A5197 due to assembly concern. Please kick off tooling modification to correct the dimension more precisely. In the meanwhile, five pieces of meeting parts were sent to your Taichung plant in accordance with your requirement. Please confirm upon the receipt of them.

    Sent with this, you may see the required sample quantity of initial PPAP and formal PPAP respectively to be submitted along with PPAP documentation. We are deeply convinced that the benefit of phased PPAP is highlighting the actual status of production run, and do appreciate you work closely with other engineering team to drive PPAP approval.

    Shall there be anything we could do for you, please

never hesitate to head up.

Yours sincerely,
Olivia Porter

 中文翻譯

衛斯理 您好：

　　特此回答貴公司昨日來電所提出之要求，我很遺憾通知您，由於裝配考量，我司工程團隊無法接受產品編號 A5197 的重點尺寸調整為 2.552。請啟動模具修改使尺寸更精確。與此同時，五件配合件已按照您的要求發送到貴公司台中廠，請確認查收。

　　如您所視，隨函覆上須與 PPAP 檔一同提交的初始 PPAP 和正式 PPAP 分別所需的樣品數量。我們深信，階段性 PPAP 的好處是突顯生產運作的實際狀況，並感激貴公司與我們工程團隊密切配合促成 PPAP 核可。

　　如果有任何需要效勞的地方，請不吝提出。

奧莉維亞波特 敬啟

翻譯小試身手

❶ 除了已排定的下列檢查項目之外，此設備還需要再進行徹底的檢查。

_____

_____

❷ 在歐盟國家裡，我們擁有最先進、最完整的文件檢查系統。

_____

_____

❸ 此次出貨的所有品項都是屬於統一關稅系統編碼 38220000 項下的產品。

_____

_____

❹ 付款交單和承兌交單都會用到國貿上常用的工具，也就是匯票。

_____

_____

參考答案   MP3 86

❶ In addition to the following scheduled examinations, the equipment will still require a thorough examination.

❷ We have the most developed and comprehensive document scrutiny system in the European Union countries.

❸ All items in this shipment are under harmonized tariff system code: 38220000.

❹ Documents Against Payment and Documents Against Acceptance both rely on an instrument widely used in international trade, i.e. a bill of exchange or draft.

Unit

11 邀請觀展 Invitation to the Trade Exhibition

國貿經驗談

▶▶ 對於公司舉足輕重的重要客戶,可在寄發邀請函前親自致電邀
約,除了可表達我方邀請觀展之誠意外,也可詢問對方預計觀展
之人員及人數中,在寄發邀請函時附上展覽入場卷。另可先預約
會議時間,我方可在展覽會場事先安排展位位置或規劃的小型會
議間進行會晤討論。

關鍵句型

at one's request 應某人要求

例句說明

· She came at my request.

她應我的請求前來。

· At your request, we offer you a special price.

應您的要求，我們提供一個優惠價。

替換句型

· Special discount is granted based on your demand.

be sure to 務必

例句說明

· Be sure to keep in touch with me.

務必與我保持聯繫。

· I'll be sure to reply within two weeks after receiving your notice.

我在接到通知後的兩周內必定回信。

替換句型

· Keep in touch with me by all means.

Issued Invitation Letter to Existing Customers with Individual Name

Dear Olivia,

It's my great pleasure, on behalf of Best International Trade Corp., to invite you to visit our booth at ISH show from March 24th to 29th, 2014.

Our business team and I will be there to handle the booth during the exhibition time, and I would be glad to show you our latest products. You may refer to the exhibition information as below, and please drop by any time.

Booth No.: 1E/23 and 1E/25
Date: March 24th ~ March 29th, 2014
Venue: Messe Frankfurt (Ludwig-Erhard-Anlage 1
        60327 Frankfurt, Hesse, Germany)

If you would like to make an appointment during the exhibit hours, please drop me a message. I can then ensure your meeting time is reserved.

Do come and join us. We anticipate to see you at the ISH Show.

Sincerely yours,
Wesley Yang

 中文翻譯

（以個人名義發送邀請函予既有客戶）

奧莉維亞 您好：

在此謹代表倍斯特公司非常榮幸的邀請您參觀找司在 ISH 展的攤位，展期為 2014 年 3 月 24 日至 29 日。

我們公司的業務團隊及本人將會在展覽期間處理攤位事宜，找將非常樂意為您展示找司最新的產品。請參閱下列展覽相關訊息，隨時歡迎您光臨。

攤位號碼： 1E/23 與 1E/35
展覽期間： 2014 年 3 月 24 日至 29 日。
展覽地點：法蘭克福展場（德國赫斯 法蘭克福，60327，路易威一鄂汗德樓1號）

如您欲在展覽期間進行會談，請與找聯絡，以確保為您事先保留會議時間。

歡迎您出席。我們十分期待在 ISH 展與您見面。

衛斯理 揚 敬啟

❶ 對於金額介於 30,000 到 100,000 歐元之間的訂單，我們所採用
的付款條件為憑單據付款。

_____

_____

❷ 若您想要選擇另一種付款條件的話，就請您在下單時告訴我們。

_____

_____

❸ 出口方若有保兌信用狀，即使外國買方或外國銀行不付款，保兌
銀行仍須支付。

_____

_____

❹ 我們可接受的付款條件為預付與不可撤銷的即期信用狀。

_____

_____

參考答案

❶ We use cash against documents as a payment term for orders between €30,000 and €100,000.

❷ If you wish to arrange alternative terms of payment, please let us know upon placing your order.

❸ An exporter with a confirmed letter of credit will be paid by the confirmed bank even if the foreign buyer or the foreign bank defaults.

❹ The terms of payment we accept are prepayment and irrevocable Letter of Credit at sight.

國貿經驗談

▶▶ 展期通常不超過一週,如何在短時間達到最高經濟效益,是參展的最主要目的,因此參展人員扮演舉足輕重的腳色,進行展前的培訓及準備工作相當重要,尤其針對首次參與展覽的新生菜鳥。培訓須著重在對產品的熟悉度,如產品規格、型號、顏色等。如能現場實際操作的產品,也須清楚產品性能及如何操作,並要瞭解各展品在攤位中的相關擺設位置。

 關鍵句型

take an interest in sth 想瞭解某事

例句說明

· My customer takes a great interest in the new material of your product.

我的顧客對於貴公司產品的新材質十分感興趣。

· Our products are well know to the buyers who take an interest in the industrial market.

我們公司的產品在熱衷工業市場買家中是眾所皆知的。

替換句型

· The client wants to learn something about your production process.

for one's reference 提供予某人參考

例句說明

· We are sending a catalogue for your reference.

我們寄了一本目錄供你方參考.

· We shall appreciate it if you will provide a pattern for our reference.

如能提供我們新花色作為參考,我司將不勝感激。

替換句型

· For more specific product information, please refer to our website.

Dear Mr. Mitsui,

"o ha yo u go za i ma su !" Please accept our thanks for your visitation to our booth during ISH show.

I'd like to know if you have had the chance to look at the catalogue offered to you at the show yet. Just to be sure, enclosed please find another catalogue.

We're pleased to have this opportunity of reminding you that the exclusive discount of exhibition will expire at the end of month. You're suggested to grasp this chance to place an initial order by the deadline.

We hope that we may be of service to you and believe you will be satisfied with our products.

Yours sincerely,
Wesley Yang

中文翻譯

三井先生 您好：

「上午好！」感謝您於 ISH 展時造訪我們公司展位。

我想知道您是否有機會看會展時提供給您的目錄。為了以防萬一，隨函附上另一份目錄。

我們很高興有這個機會提醒您，展覽的獨家折扣將在本月底到期。建議你抓住這個機會在截止日前下首張訂單。

希望我們公司可以為您服務，相信您會滿意我們公司的產品。

衛斯理 揚 敬啟

❶ 你知道可轉讓的票據文件是可以兌現的嗎？

_____

_____

❷ 在這些帳款尚未付清前，我們將無法執行任何訂單，而對於遲付
貨款，我們也保有收取利息的權利。

_____

_____

❸ 能否請您告知您用來付款的那張信用卡的三碼安全碼？

_____

_____

❹ 若是您對您的發票有任何問題，請撥打我們的應付帳款諮詢專
線。

_____

_____

參考答案　　　　　　　　　MP3 88

❶ Do you know negotiable documents can be exchanged for money?

❷ We will not be able to fulfill any further purchase orders while these accounts remain unpaid and reserve the right to add interest on to late payments.

❸ Could we please have the 3-digit security code for the credit card you are using for payment?

❹ If you have any queries regarding your invoices, please call the Accounts Payable Helpline.

Unit 13

展後拜訪 Visitation Plan to Potential Customer Meeting in the Exhibition

國貿經驗談

▶▶ 綜合上述,參展行程約為期十到十四天,返回工作崗位後,除了補足這兩周來未處理或未完整處理的庶務性工作外,另一個重要的工作就是整理展覽紀錄,排出優先處理順序,進行後續追蹤。完成答應客戶的代辦事項應列在首要,如回覆詢價或提供樣品等。而針對僅索取目錄,無特定代辦事項的客戶,可列在最後,追蹤對方檢視目錄之結果及採購意願針對。而針對可能啟動交易的目標客戶,可另外安排拜訪行程,積極爭取合作機會。

關鍵句型

be convenient to (for) sb 對某人是方便的

例句說明

· Call me at any time that is convenient to you.

任何您方便的時候就來電。

· It is not convenient for me to ring her up.

我現在不便於致電給她。

替換句型

· We will call her at the proper time.

· It is not the proper time for me to ring her up.

fly to some place on business 到某地出差

例句說明

· Annie flied to Tokyo on business.

安妮到東京出差。

· Did you fly to New York on business or for pleasure last month?

你上個月到紐約是洽公還是遊玩？

替換句型

· Annie went on a business trip to Tokyo.

· Annie was sent to Tokyo on a business trip.

· Annie was away on a business trip to Tokyo.

Dear Mr. Douglas,

We were pleased to see you during the ISH show last month and it's a pleasure to learn you're interested in our 3/4 ball valve.

I'm going to be on a business trip to your city in week 16 and wondering if you will be available to meet me to talk about further cooperation plan.

Please let me know if the arrangement is convenient for you. If not, please advise when the most convenient time for my visit is.

Looking forward to your early reply.

Yours sincerely,
Wesley Yang

中文翻譯

道格拉斯 先生 您好，

　　我們很高興上個月能在 ISH 展見到您，以及知道貴公司對我們的 3/4 英吋球閥感興趣。

　　我將在第十六週造訪貴公司所在城市，屆時不知您是否方便與我會面談談未來合作計畫。

　　請讓我知道這項安排對您來說是否方便。若否，請告知何時拜訪您最方便。

　　期待你的早日答覆。

　　　　　　　　　　　　　　　　　衛斯理 楊 敬啟

❶ 請跟我們的應收帳款部門聯絡,問問新銀行帳戶的資料。

❷ 我們銀行帳戶資料自 2017 年 1 月 18 日起已更改,請您確認您的匯款會匯至正確的帳戶。

❸ 我們剛匯款到您的帳戶,但銀行說您們所給的銀行國際代碼不正確,請盡快回覆,好讓我們能完成付款。

❹ 國際銀行的匯票是寄錢給其他國家的人的一種安全的方式。

參考答案   MP3 89

❶ Please contact our Account Receivable Department for new bank account information.

❷ As of the 18th of January 2017, our BANK Account details have changed. Please make sure to send your remittance to the correct account.

❸ We just wired the payment to your account but our bank said the given SWIFT code for your bank is incorrect. Please reply asap enabling us to complete the payment.

❹ An international bank draft is a secure way to send money to someone in another country.

Unit

14 簽證申請
Applying for Visa

 國貿經驗談

▶▶ 外籍人士申請來台洽商停留簽證除須規定有效期為 6 個月以上護
照、簽證申請表及照片等外，另需提供台灣境內商務合作夥伴所
出具的商務活動證明檔，通常是出具邀請函件，該邀請函須包含
被邀請者個人資訊（姓名、性別、出生日期等）、訪問資訊（入
境事由、訪問期間、訪問地點、邀請單位、費用歸屬等）、以及
邀請單位資訊（邀請單位名稱、電話、地址、單位用印、法定代
表及其簽章等）。另需特別注意被邀請人個人資料需與護照所列
一致。

 關鍵句型

Please grant sb. sth. 請允許某人某事物

例句說明

· Please grant him his request.

請答應他的請求。

· Please grant me the formal authorization for the Germany trip.

請正式批准我出差德國。

替換句型

· Please formally authorize me for the Germany trip.

Sth. will be borne by sb. 某事物費用由某人承擔

例句說明

· The return shipping will be borne by the seller.

退貨費用將由賣方承擔。

· The exhibition fee and all relevant expenses will be borne by the joint exhibitors on an equal footing.

· 參展費用及所有相關花費將由聯合展商平均負擔。

替換句型

· The return shipping will be at seller's expense.

Main Office - Best Corp. in Shanghai, Mainland China
1 Nanjing Road East Shanghai, Shanghai 200001, China
Tel: 86-21-1111-1111 / Email: best.shoffice@best.com

INVITATION LETTER

To whom it may concern:

It's our great honor to invite Ms. Olivia Porter of ABC Co., Ltd. to visit the main office of Best Group in Shanghai to discuss our future cooperation plan. Her business trip will start from March 1st, 2015 to March 14th, 2015. To strengthen and develop the trade relations between both sides in the future, Ms. Olivia Porter will come to China many times. Please grant Ms. Olivia Porter the necessary privileges to obtain a Multiple Visa to visit China. Her personal details is listed as below:

Name: Olivia Porter / Date of birth: 1st JAN, 1969
Citizenship: the United States of America (U.S.A.) / Passport #: C000111

Please note that during Ms. Poter's stay in China, Best Group does not assume any legal or financial responsibility. All relevant expenses of her visitation will be also borne by her company.

We greatly appreciate your assistance in prompt processing the necessary document to Ms. Porter.

Yours sincerely,
Best Group
1ˢᵗ Jan, 2015

 中文翻譯

辦公室－倍斯特公司，位於中國上海

中國上海，上海南京東路 1 號，郵遞區號 200001

電話：86-21-1111-1111 | 電郵：best.shoffice@best.com

邀請函

敬啟者：

　　我們很榮幸地邀請 ABC 有限公司的奧利維亞 波特小姐參觀貝斯特集團在上海的主要辦公室，討論雙方未來合作計畫。她出差期間將從西元二零一五年三月一日開始至西元二零一五年三月十四日止。為鞏固和發展雙方未來貿易關係，奧利維亞 波特小姐將會多次來訪中國。請給予必要的權限，賦予奧利維亞波特小姐訪問中國的多次簽證。她的個人資訊如下：

姓　　名：奧利維亞 波特

出生日期：西元一九六九年一月一日

國　　籍：美國

護照號碼：C000111

請注意，波特小姐在中國停留期間，倍斯特集團不承擔任何法律和財務責任。她訪問期間的所有相關的費用也將由她所屬公司承擔。

我們非常感謝您的協助及時辦理波特小姐所需的文件。

倍斯特集團 敬啟
西元二零一五年一月一日

翻譯小試身手

❶ 付款人通常是進口人或是代收銀行。

❷ 我們可以開立信用狀，給您的海外供應商提供個安全的付款方式。

❸ 除非另有協議，否則是允許分批出貨的。

❹ 若是禁止轉運，可能會有問題發生，因為現今的運輸複雜多了。

參考答案　　　　　　　　　　　　　　MP3 90

❶ The drawee is generally the importer or the collecting bank.

❷ We can issue a Letter of Credit so as to provide a secure means of payment to your overseas suppliers.

❸ Partial shipments shall be permitted unless otherwise agreed.

❹ Problems may arise if transshipment is prohibited, as transport is a complex business these days.

15 交通安排 Transportation Arrangement

 國貿經驗談

▶▶ 表達感謝的語句口語上多數以「Thank」，及「Appreciate」居多，需注意的一點是「thank」當作動詞 (v.) 時，其後接感謝的對象，而後加上介係詞 for 及感謝的事項，即 thank + sb. + for sth.，例句：*We thank you for your assistance.*；而「appreciate」當作動詞 (v.) 時，其後可直接加上 sb. 或 sth.，例句：*We appreciate your assistance* 或 *We appreciate you.*

關鍵句型

A be in contact with B　A 與 B 聯繫關於

例句說明

· The buyer is in daily video contact with the seller to follow up the production progress.

買方每天視訊聯繫賣方追蹤生產進度。

· Jack is always in direct contact with the customer by phone call about shipping schedule.

傑克通常直接電話聯繫客戶出貨事宜。

替換句型

· Jack always communicates with the customer by phone call directly about shipping schedule.

Sb. take care of sth. on one's own　某人自行處理某事

例句說明

· You have to take care of the shipping schedule on your own.

你必須自行處理出貨排程。

· The manufacturers must take care of the product inspection on their own before releasing.

製造商在出貨前須自行安排產品檢驗。

替換句型

· You have to deal with the shipping schedule by yourself.

Dear Olivia,

    I hereby acknowledge receipt of your itinerary and well noted that <u>you will take care of the hotel reservation on your own.</u>

    Per your requirement, please refer to the suggested timetable from Taoyuan to Taichung as below. Or, you can go to THSR website on http://www5.thsrc.com.tw/en/ for more details.

Train I.D.	Remark	Departure Time	Arrival Time	Adult	Children Senior Disabled	Group Ticket
763		22:15	22:52	NT$590	NT$295	NT$560
399		22:31	23:01	NT$590	NT$295	NT$560
541		22:52	23:30	NT$590	NT$295	NT$560
545		23:22	23:59	NT$590	NT$295	NT$560

    Shall there be anything I can do for you, please never hesitate to contact me. I look forward to seeing you soon.

Yours sincerely,
Wesley Yang

中文翻譯

奧利維亞 您好：

特此告知收到您的行程，並已知悉您將自行處理酒店預訂事宜。

依據您的要求，請參考建議的桃園至台中之班次時間表如下。或者，您可以直接上高鐵網站 http://www5.thsrc.com.tw/en/ 獲知詳情。

班次	備註	出發時間	抵達時間	全票	半票優待票	團體票
763		22:15	22:52	NT$590	NT$295	NT$560
399		22:31	23:01	NT$590	NT$295	NT$560
541		22:52	23:30	NT$590	NT$295	NT$560
545		23:22	23:59	NT$590	NT$295	NT$560

如有任何我司可以為您效勞的地方，請不要客氣聯絡我。期待盡速見到您。

衛斯理揚 敬啟

翻譯小試身手

❶ 若所出貨品的數量誤差值在所訂數量的上下 5％以內的話，那將視此供應商已履行了合約。

_____

_____

❷ 我們會對遲繳的貨款收取利息，以彌補催收欠款所產生的成本。

_____

_____

❸ 請確定您們都有收到這些發票了，也請確保待這些發票到期時，您們會依我們所議定的付款條件來支付貨款。

_____

_____

❹ 我們的紀錄顯示您單號 105234 的這份訂單仍有餘額未清，請查一查，並告知我們何時可收到貨款。

_____

_____

參考答案   MP3 91

1 The supplier shall be deemed to have fulfilled its contract if the goods delivered fell within this tolerance of plus or minus five percent (5%) of the quantity ordered.

2 We'll charge late payment interest to help recover the cost of debt collection.

3 Please ensure that all of these invoices have been received by yourselves and when due, they are to be paid in line with our agreed payment terms.

4 Our records show an outstanding balance for your order 105234. Please check and let us know when we can expect the payment.

Unit 16 食宿預定 Restaurant & Hotel Reservation

 國貿經驗談

▶▶ 協助客戶訂房後可將飯店的訂房確認書或訂房號碼提供予客戶，以便加快飯店辦理客戶入住登記時之查詢流程，使客戶在經歷舟車勞頓後，能順利且快速地下榻飯店休息。

關鍵句型

Sb. be delighted 某人非常樂意

例句說明

· I'd be absolutely delighted to attend your anniversary party.

我非常樂意參加您的週年派對。

· We are very much delighted with the news of sample approval.

我們非常高興得知送樣通過。

替換句型

· To our delight, the sample is approved.

A meet B at/in some place A 與 B 在某地會面

例句說明

· Shall I meet you in the hotel lobby directly?

我可以直接在飯店大廳接妳嗎?

· I tend to meet you in the hotel restaurant to discuss the project details over the lunch.

我想與你在飯店餐廳會面,在席間邊吃邊談專案細節。

替換句型

· Shall I go to the hotel lobby to meet you directly?

Dear David,

    <u>I were delighted to</u> receive your itinerary and have arranged the hotel reservation according to your requirements as below. Please also find attached hotel reservation letters.

Date	Jan. 28th	Jan. 29th ~ Feb. 1st
Hotel	Sheraton Grande (in Taipei)	Windsor Hotel (in Taichung)
Room Type	Premier Room / King Bed * no smoking * w/ breakfast	Deluxe Room / King Bed * no smoking * w/ breakfast

    We will send a van to pick you up from Tao Yuan Airport to Sheraton Grande Taipei Hotel on Jan. 28th. Next morning our CEO and <u>I will meet you personally</u> on 9:00am at the hotel and organize a welcome dinner for you that night. On the last day, said Feb. 2nd, the van will drive you to the airport by 6:00am. Please feel free to let me know for any need during your stay. I look forward to seeing both of you.

Yours sincerely,
Wesley Yang

 中文翻譯

大偉 您好：

　　很高興收到您的行程，並已根據您的要求預訂酒店如下。同時附上附件訂房確認書。

日期	一月二十八日	一月二十九日 至 二月一日
飯店	臺北喜來登酒店	台中裕元花園酒店
房型	行政客房/ 加大床 ＊非吸菸房 ＊含早餐	豪華客房 / 加大床 ＊非吸菸房 ＊含早餐

　　我司將派車接您們從桃園國際機場到臺北喜來登酒店。第二天早上，我司執行長和我將親自在早上九點到飯店接您們，並於當天晚上為您們舉辦歡迎晚宴。最後一天，即二月二日，我們的車會在早上六點前開車送您們到機場。在您們停留期間如有任何需要，請隨時讓我知道。期待見到二位。

　　　　　　　　　　　　　　　　　　　　　　　衛斯理揚 敬啟

❶ 我們的保固責任始於新品購買日當天。

❷ 這個部分包含了問題解決的資料，應有助您解決大部分的問題。

❸ 若是沒有遵照保固條款的規定，包括像是產品安裝不當，那麼保固將會失效。

❹ 若符合我們的退貨政策，我們就會在收貨後 14 天內退款給您。

參考答案 MP3 92

❶ Our warranty obligation starts on the date of the purchase of a new product.

❷ This section covers troubleshooting information which should help you solve the majority of problems.

❸ The warranty will be invalidated if any of the terms and conditions of the warranty are not adhered to, including incorrect installation of the product.

❹ In compliance with our return policy, we will issue a refund to you within 14 days of receipt.

Unit
17 人員介紹 Introduction of Company Staff

💡 **國貿經驗談**

▶▶ 不論是客戶來訪或外出洽公拜訪客戶，務必要事先使用專用名片夾準備好個人名片，以利在自我介紹時能及時取出名片，避免臨時找不到名片的尷尬場面，而遞交名片時由位階較低至位階較高的順序遞交，雙手遞交名片後再自我介紹個人姓名、所屬部門及職稱。

關鍵句型

Sb. be grateful for… 感謝某事

例句說明

・I am grateful for your advice.

感謝您的忠告。

・I should be grateful if you would send out the sample at an early date.

如能早日寄出樣品，將不勝感激。

替換句型

・I am indebted for your advice.

・I am obliged for your advice.

・I am thankful for your advice.

Dear Wesley,

I hereby acknowledge receipt of the invitation letter of your year-end party. It's truly a pleasure for me to attend the party.

My plan is stay in your city for 3 days and spends one day to visit your office for new project discussion. If time permits, I'd also like to tour your plant. The itinerary of my stay will be shared once finalized.

Looking forward to seeing you and your team members in person.

Yours sincerely,
David Brad

中文翻譯

衛斯理 您好：

特此告知敬收貴公司年終派對邀請函。能參加派對實屬榮幸。

我計畫在貴城市停留三日，一日時間拜訪公司辦公室討論新專案。如果時間許可，我還想參觀貴司工廠。待行程敲定後，將提供予您。

期待親自見到您及您的團隊成員。

大衛布萊德 敬啟

翻譯小試身手

❶ 請見此 e-mail 所附的號碼 SI1202 這一張發票，它已逾期，尚未付款。

❷ 若是仍不結清帳款，我們將有權對遲繳的貨款加計利息。

❸ 所有帳款應於發票日起 30 日內付清。

❹ 我們稽核的程序之一是要確認 2016 年 12 月 31 日的應收帳款餘額。

參考答案　　　　　　　　　　　　　　　MP3 93

❶ Please see attached to this e-mail a copy of our invoice # SI1202 which remains outstanding and overdue for payment.

❷ If accounts remain unpaid, we reserve the right to add interest on late payments.

❸ All accounts are to be settled within 30 days from the date of invoice.

❹ As part of our audit procedures, we would like to confirm the accounts receivable balances as of December 31, 2016.

Unit

18 公司介紹
Tour of Office Building

國貿經驗談

▶▶ 國外客戶參訪，若為首次來訪客戶，我方接待人員可先帶領客戶熟悉辦公室環境，並在環境介紹過程中順勢介紹各部門主管，唯介紹業務部門時可一一向客戶介紹相關負責業務組員。環境及人員介紹完畢後，則引領客戶至會議室稍坐片刻，接著可由專人進行公司簡介或欣賞公司簡介影片。最後即進入主題，即會議章程討論。

 關鍵句型

Sth. be located behind 某物坐落於…之後

例句說明

· The new exhibition hall is just located behind the old one.

新展示廳就坐落在舊展示廳後方。

· The restroom is located behind the meeting room.

洗手間位於會議室後方。

替換句型

· The new exhibition hall is set just behind the old one.

Let me lead sb. to 讓我引領某人至

例句說明

· Let me lead you to the new exhibition hall.

讓我引領你到新展示廳。

· Let me lead you the way.

我替你帶路。

替換句型

· Let me guide you to the new exhibition hall.

Dear Mr. Brad,

How's everything? It's so glad to learn that you will have a business trip to our country from March 20 to 27. Is it possible for you to arrange an appointment to discuss our initial cooperation plan during your stay here? Besides, I would like to take this opportunity to show you our new-built office building.

If this arrangement is workable, please let me know the potential date being convenient for you.

We look forward to your confirmation soon.

Yours sincerely,
Wesley Yang

中文翻譯

布萊德先生 您好：

　　一切都好嗎？很高興得知你將於三月二十日至二十七日期間出差至我國。在您停留本地期間是否可能安排會面討論我們的初步合作計畫？另外，我想藉此機會帶您參觀我們的新落成的辦公樓。

　　如果這個安排可行，請讓我知道您方便的日期。

　　期盼您盡速確認。

衛斯理揚 敬啟

❶ 此對帳單並不是要來催收帳款,只是為了稽核之用。您在 2016 年 6 月 30 日後所付的款項並不會反映在結餘裡。

_____

_____

❷ 貸項通知單是會計上的一個方法,用來減少發票金額或取消發票。

_____

_____

❸ 我們價格調整了 3%,再加上銷售額增加了 10%,才又將我們回復到財務平衡的狀態。

_____

_____

❹ 請在發票日後 30 天付款,付款後也請告訴我們,發匯款通知單給我們。

_____

_____

❶ This is not a request for payment but for audit purposes only. Payment received after June 30, 2016 is not reflected in the balance shown.

❷ A Credit Memo is an accounting method of reducing or canceling an invoice.

❸ Our 3% price increase as well as a 10% increase in the number of sales has brought us back to financial equilibrium.

❹ Please make payment within 30 days from the invoice date and to notify us with a Remittance Advice once payment has been made.

19 參觀工廠 Tour of Facility

國貿經驗談

▶▶ 產品依生產完成度主要區分為成品、半成品、在製品。成品是指已完成全部生產製程,且檢驗合格並入庫,可供銷售的產品;半成品,是指在單一車間已完成加工,且經檢驗合格後入半成品倉,待轉入下一車間繼續加工的產品。在製品,是指仍在單一車間的工序上,正在加工的製品,或雖已加工完畢但尚未檢驗入半成品倉或成品倉的製品。因此在製品是介於原材料和半成品之間,半成品和半成品之間,以及半成品和成品之間的製品。

關鍵句型

on one's right / left 在某人右手/左手邊

例句說明

· The showroom is about two meters away on your right.
展示間就在你右手邊約兩公尺遠的地方。

· My meeting room is on your left.
我的辦公室在你左手邊。

替換句型

· The showroom is about two meters away to your right side.

Sb. concentrate on 某人致力於

例句說明

· We concentrate on high quality of product and service.
我們致力於提供高品質產品及服務。

· Most of the manufactures concentrate on cost reduction.
多數製造生致力於降低成本。

替換句型

· We focus on high quality of product and service.

Dear Olivia,

I am very pleased to hear that you will visit Taiwan next Monday. Unfortunately, I will be on a business trip to Hong Kong on that day but will be back that evening.

Are you available to stop by Taichung City Monday afternoon? David Chen, our senior plant director, will be very pleased to take this opportunity to be your tour guide of our Taichung facility. I can meet you to have a cup of coffee at your hotel around 8:00P.M if it's convenient for you.

Please confirm by return if this arrangement is fine for you.

Yours sincerely,
Wesley Yang

中文翻譯

奧莉維亞 您好：

很高興得知您下星期一要造訪臺灣的消息。不巧的是，那天我會出差香港，但當天晚上即會返台。

不知您是否能在星期一下午停留台中？我們的資深廠長大衛 陳非常高興能藉此機會做為您參觀我們台中廠的響導。如果您方便的話，我可以約在晚上八點在您下榻的酒店與您會面喝一杯咖啡。

請回復確認這個安排對您來說是否可行。

衛斯理 楊 敬啟

❶ 在此附上您的對帳單,您的帳上金額已超出信用額度,還請您盡早付款,這樣之後的出貨才不會受到延誤。

❷ 在此我要宣布一下我們有推出了幾份新型錄,這些型錄也已加進網頁的行銷素材資料裡,可讓您發送給客戶。

❸ 我們很難在非標準規格的相似產品之間做比較。

❹ 在此附上我們建議的單張型錄讓您看看,也讓您在即將到來的會議上使用。

參考答案 MP3 95

❶ Your statement is attached. You have exceeded your credit limit. Please remit payment at your earliest convenience, so future shipments will not be delayed.

❷ I would like to announce several updates regarding our brochures. They're added on our marketing material webpage and available for distribution.

❸ It can be difficult for us to make comparisons between similar products which do not come in standard sizes.

❹ I've attached our recommended flyers for your review and use at the upcoming conference.

Unit

20 認證稽核 Certification

國貿經驗談

▶▶ 歐美各國客戶十分注重認證（Certification），不論是對產品本身、生產製造流程、製造商人員、生產單位，甚至是出口商及其相關生產業者的社會責任等等。因此在與出口商及其相關生產業者合作前後即可能提出認證需求。依據各產業別不同，所需認證的項目及機構也有所差異，一般來說是由國外客戶向認證機機構提出需求，由出口商及其相關生產業者填具認證機構所需的相關文件後，由國外客戶將相關文件提交予認證機構，而後由認證機構派員至出口商及其相關生產業者進行認證審核，最終依認證合格與否提供稽核報告（Inspection Report），合格者另發給證書（Certificate）。

 關鍵句型

as detailed in the attached (document) 如附（文）詳載

例句說明
- The inquired piece prices are as detailed in the attached sheet.
  所需產品單價如附件詳載。
- You may refer to our working calendar of next year as detailed in the schedule table.
  請參閱我司明年度行事曆，詳細如附件行程表。

替換句型
- The inquired piece prices are as particularized on the enclosed sheet.

Permit me (us) to remind sb. …. 容我（我們）提醒某人

例句說明
- Permit me to remind the new staff that training will start after lunch.
  容我提醒新人，教育順練在午餐後開始。
- Please permit us to remind you that our office will be closed for 5-day-consecutive Christmas holiday starting Dec. 23.
  請容我司提醒您，我司辦公室將從十二月二十三日起連休五天聖誕假。

替換句型
- Please allow us to call your attention that our office will be closed for 5-day-consecutive Christmas holiday starting Dec. 23.

Dear Wesley,

We enclose herewith spread sheet for IAPMO Lead Free Certification. Please fill out the areas highlighted in yellow completely for each size on a different sheet. As detailed in the attached documents, what we suggested is listing just the major components including the body halves, the ball valve, the stem assembly, and other related mating parts contacting water surface, instead of whole kit. Even though some of these components do not contain lead, but contact water, they still need to be documented.

Permit us to remind you that the required documents should be sent by return by next Monday. Any question, please feel free to let me know.

Yours sincerely,
Olivia Porter

中文翻譯

衛斯理 您好：

　　隨函附上 IAPMO 無鉛認證表格。請將個別不同尺寸產品的資料完整填寫在各個頁籤的黃色標記區域。如附檔詳載，我司的建議是只列出主要部件包括閥體，閥球，閥杆組件，以及其它會與水接觸的相關配合件，而非完整組裝套件。儘管部分所列元件不含鉛，但會接觸水，因此仍然需要被列表。

　　容我們提醒您，需在下週一前寄回所需的文件。如有任何問題，請隨時讓我知道。

奧利維亞 波特 敬啟

❶ 請看仿單資料,裡頭有列出所有的成份、警告、注意事項、副作用、臨床結果,以及其他重要的醫藥資訊。

_____

_____

❷ 請記得透過我們內部的文宣網頁來提出您要求的行銷素材,並完整填寫申請表格。

_____

_____

❸ 為了確保有最佳的品質表現,請在使用此儀器前先閱讀這份操作手冊。

_____

_____

❹ 若您可給予任何的回饋訊息,我們絕對會非常感激,我們也隨時可以回答您的任何問題。

_____

_____

參考答案 MP3 96

❶ Please see the package insert for the complete list of indications, warnings, precautions, adverse events, clinical results, and other important medical information.

❷ Please remember to send all your marketing material requests through our internal literature page and fill out the form completely.

❸ To ensure the best performance, please read the Operating Manual before using the instrument.

❹ We would certainly appreciate any feedback from you and are available to answer any of your questions at any time.

國貿經驗談

▶▶ 尾牙通常會在農曆新年前的一個月內前後舉辦，此時正值歐美地
區聖誕假期後所迎接的新第一季度，也是客戶擁有充足出差預算
之開始，因此可抓緊機會力邀客戶參加尾牙宴，藉此增進彼此關
係，爭取新年度合作方案及訂單。

關鍵句型

sb. / sth. just happen to 某人/某事正巧是

例句說明

· Nina just happened to meet me in the exhibition last week.

妮娜上週在秀展巧遇我。

· Nina's traveling date just happens to be the same day as mine.

妮娜的出差日與我的出差日恰巧是同一天。

替換句型

· Nina just chanced to meet me in the exhibition last week.

Is it necessary ….. ? 是否需要….

例句說明

· Is it necessary to arrange airport transfers for you?

是否需要為您安排機場接送?

· Is it more necessary to work an extra shift on Sunday during the peak times of production?

是否更需要在訂單高峰期安排周日加班?

替換句型

· Do you need any assistance in arranging airport transfers?

Dear Olivia,

It's my pleasure, on behalf of Best Group to invite you to our company's year-end dinner banquet at Formosa Hotel from 6:00PM to 10:00PM on Friday, Feb. 1st.

Best Group would like to take this opportunity to appreciate your great support to achieve roaring success on business during the year. It would be our great honor if you could attend the party with your family. For your convenience, with attached are a map and detailed directions of Formosa Hotel.

Best Group sincerely hopes you can join us at the year-end dinner banquet. Please RSVP if you can come by January 1st, so that we can ensure your place is reserved.

Yours sincerely,
Wesley Yang

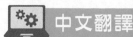
中文翻譯

奧莉維亞 您好：

　　謹代表倍斯特集團，很榮幸地邀請您參加我司在二月一日晚上 6:00 到 10:00 於福爾摩沙酒店舉辦的年終尾牙晚宴。

　　倍斯特集團想借此機會感謝您的大力支持，致使我們今年度業務大有斬獲。如您能與您的家人參加晚宴，將是我們的榮幸。為了您的方便，附件為福爾摩沙酒店的地圖及詳細方位。

　　倍斯特集團真誠地希望您能參加我們的年終宴會。請在一月一日前回覆您是否能前來，以便保留您的座位。

衛斯理 楊 敬啟

❶ 點入這裡就可看到所有在 2016 年十二月所推出的新產品

❷ 謝謝您提供客戶與競爭者的資訊,不過,因為生產成本高,我們沒辦法配合提供跟競爭者相同的價格。

❸ 我們可對此產品在使用者收貨日後一年內的品質表現提供保證。

❹ 這個案子的大多數產品我們給到 60%的折扣,應能彌補在新產品部分的 25%折扣限制,讓我們的價格還能有競爭性。

參考答案   MP3 97

1 Click here to view all New Products launched in December 2016.

2 Thank you for providing customer and competitor information. However, due to high production costs, we are not able to match competitor's pricing.

3 We guarantee the performance of this product for one year from the date of receipt by the end user.

4 The 60% discount granted for most products in this project should make up for the 25% discount restriction on the new products, and our price could still be competitive.

Unit

22 人事異動 Reshuffle Announcement

國貿經驗談

▶▶ 就如同人員請假一樣，如逢人事異動時，人事單位或相關業務單位之主管應即時主動發函通知客戶，除說明異動原因外，主要是告知新的聯繫窗口（contact window），包含承接人員姓名、職稱、電話（分機）及電郵信箱等。尤其對國外客戶而言，無法直接至供應商端了解業務，因此透過電郵及電話聯繫特定業務人員是最直接的方式，所以客戶端通常不太樂見人員異動之情形，主要仍是擔心新承辦人員不熟悉及了解業務狀況。對於分工較細之企業，可提供聯繫窗口清單，包含各項業務負責人員，直屬主管及部門最高主管相關聯繫資料。

🔍 關鍵句型

take a good look at 仔細看一下

例句說明

· Let's take a good look at earnings report of this month.
讓我們仔細看一下本月的財報。

· You'd better take a good look at the quotation sheet before offering.
在報價前,你最好仔細看清楚。

替換句型

· You'd better look at the quotation sheet carefully before offering.

I have heard a lot about you. 久仰大名

例句說明

· I have heard a lot about you. Pleased to meet you.
久仰大名,請多指教!

替換句型

· I have heard so much about you. Nice to meet you.

· I have heard a great deal about you. Glad to see you.

Dear Customers,

This is to inform you that Mr. Tony Yang has been promoted to general manager of the Sales Department. We'd like assign Mr. Wesley Yang to take over his position and be responsible for the trade issues between us.

Mr. Wesley Yang has been with Best Group more than ten years. He works seriously and tidy at original understanding and idea. We firmly believe his excellent ability and personal qualities will make great contributions to our future cooperation and consolidation of good trade relationship between two parties.

We would like to express our warmest thanks for your continued support throughout the years. Shall you have further questions and demands, please feel free to contact Wesley via wesleyyang@best.com.

Yours sincerely,
Best Group

中文翻譯

親愛的顧客 您好,

　　茲以本文通知您,湯尼 楊先生已晉升為銷售部總經理。我們公司將委任衛斯理 楊先生接替他的職務,負責我們的貿易議題。

　　衛斯理 楊先生已在倍斯特集團服務逾十年。他工作嚴謹有條理,且見解獨到。我們堅信他出色的能力和個人特質,將對我們未來的合作與鞏固雙方良好貿易關係,作出巨大貢獻。

　　我們由衷感謝貴公司多年來的持續支持。有進一步問題及需求,請隨時透過電郵 wesleyyang@best.com 與衛斯理聯繫。

倍斯特集團 敬啟

❶ 附上出口許可證供您留做紀錄，其效期至 2019 年 6 月 30 日。

_____

_____

❷ 這些文件得要經由公證人辦理公證，予以證實，這樣我們的主管機關才會接受這些文件。

_____

_____

❸ 必須要有您的商業登記證影本，才能完成註冊程序。

_____

_____

❹ 請在貨到後檢查一下，若有任何短缺、瑕疵或損壞的情況，請以書面方式通知我們。

_____

_____

參考答案   MP3 98

❶ The Export License is attached for your records. It is valid through June 30, 2019.

❷ You have to have the documents notarized by a Notary to validate them, so they will be accepted by our authorities.

❸ A copy of your business registration certificate will be required to complete the registration process.

❹ Please inspect the product on arrival and notify us in writing of any claims for shortages, defects, or damages.

23 喬遷啟示
Removal Notification

 國貿經驗談

▶▶ 目前與國外客戶訊息往來多已電子郵件居多,但仍會有需要藉由郵遞提供文件、樣品等其它資訊的情形。由我方寄件至國外,直接標註英文住址即可。但若由國外寄件至我方,則建議我方提供對方中、英文並列之住址,客戶可同時將中英文住址標註在收件人欄位,不僅可避免我方郵局或快遞公司因英譯差異導致誤送,亦可減少遞送之時間。

關鍵句型

be on behalf of sb. 代表某人

例句說明

· I'm writing on behalf of my supervisor to express his appreciation.

我代表上司寫此文表達對您的感激之情。

· Joe spoke on behalf of her colleagues to demand higher pay.

喬代表同僚發言要求加薪。

替換句型

· Joe spoke in the name of her colleagues to demand higher pay.

A tell B to make sure… A 要求 B 確定某事

例句說明

· The manager told her assistance to make sure you'll be present at the meeting.

經理要求助理在下班前確定您會與會。

· The buyer told the supplier to make sure of the punctual shipment.

買方要求供應商確保準時出貨。

替換句型

· The supplier was asked to insure the punctual shipment by the buyer.

Dear Customers,

    We're glad to announce that Best Headquarter will move to a new-built Formosa building with address and contact information as below. The new office will be in operation from May 01, 2015.

Headquarter - Best International Trade Corp. in Taipei Taiwan
Address: 20F.-G, No.100, Fuzhou St., Zhongzheng Dist., Taipei City,
10078 Taiwan, R.O.C.
Tel: 886-2-2351-6666
Fax: 886-4-2351-8888
Email: best.headquarter@best.com.tw

    To express our immense gratitude of your continued support and cooperation for these years, we plan to hold afternoon-tea buffet at new office building from 2:00PM to 5:00PM on May 15.

    The whole staff of Best sincerely look forward to your presence.

Yours sincerely,
Best Group

中文翻譯

親愛的客戶 您好：

很高興的宣佈倍斯特總部即將遷移到新落成的福爾摩沙大廈，位址及聯絡資訊如下所列。新辦事處將於 2015 年 5 月 1 日啟用。

集團總部 -倍斯特國際貿易公司 / 臺灣臺北
10078 臺灣臺北市中正區福州街 100 號 20 樓之 G
電話：886-2-2351-6666
傳真：886-4-2351-8888
電郵：best.headquarter@best.com.tw

為表達我司對您這些年持續支持與合作的無限感激，我司計畫在五月十五日下午二點至五點於新辦公樓舉行自助下午茶會。

倍斯特全體同仁真誠期待您的蒞臨。

倍斯特集團 敬啟

❶ 在應用產品之前，使用者應決定此產品是否適合其預定的用途。

❷ 我們需要一份國家主管機關所出具的「無異議信函」。

❸ 等您一收到進口許可證，就必須馬上寄影本給我們，因為 DEFRA
證明裡頭的用語得要跟您們進口許可證所寫的一樣。

❹ 這些商業文件得要經由商會認證。

參考答案

MP3 99

❶ Before utilizing the product, the user should determine the suitability of the product for its intended use.

❷ We need a "Letter of no objection" issued by the appropriate competent national authority.

❸ Once you receive your import permit you need to send us a copy because the wording on the DEFRA[2] Certificate must be the same as on your import permit.

❹ These commercial documents may need to be authenticated by a Chamber of Commerce.

Unit

24 表達祝賀 Expressing Congratulation

國貿經驗談

▶▶ 多數西方國家客戶對於具中國風的文物都相當感興趣,例如仿古古董文物、琉璃、玉飾等等。因此只要是東方色彩濃厚之禮品,都頗受西方客戶青睞。然而東風色彩文物多多少少會帶有宗教特色在,例如佛尊玉飾,因此宗教信仰仍是送禮時的必要考量之一,畢竟西方國家仍以信仰基督教居多。

關鍵句型

Sb. does not expect… 某人不預期…

(例句說明)

· The manufacturer does not expect the price increase of raw material.

製造商並無預期原物料價格上漲。

· I don't expect to get the business.

我並不預期談成這筆生意。

(替換句型)

· I'm not supposed to get the business.

Sb. wouldn't be surprised… 某人對….不會感到意外

(例句說明)

· The manufacturer wouldn't be surprised for the price increase of raw material.

製造商對於原物料價格上漲並不會感到意外。

· I wouldn't be surprised to get the business.

我對於談成這筆生意並不會感到意外。

(替換句型)

· The price increase of raw material wouldn't take manufacturer by surprise.

Dear Tony,

On behalf of ABC Co., I'd like to congratulate you on your promotion to the position of the general manager in the Sales Department.

Throughout recent years, you have expended a lot of effort in your company. It's your great talent leading to amazing performance of Sales Department. All your hard work has not gone unnoticed. It's for sure that the department will become more prosperous under your management.

I wish you every success, and hope our cooperation will create prosperity magnificently in the future.

Yours sincerely,
Tom Smith

中文翻譯

湯尼 您好：

僅代表 ABC 公司祝賀您榮升為銷售部總經理。

這幾年您為公司投入了大量的心力。因為您的卓越才能，使銷售部門能有驚人的業績表現。您所有的努力沒有白費。可以肯定的是，在您的管理下，部門將更加繁榮。

祝您成功，並希望我們今後的合作會創造輝煌榮景。

湯姆 史密斯 敬啟

❶ 買方若發現有任何瑕疵，應立即以書面方式通知賣方，而買方可將瑕疵品退回賣方，所有相關成本須先由買方預付。

❷ 從現在開始，請使用這些新版的問卷表格，舊版表格就請捨棄不用。

❸ 我們保證會讓您對我們的產品有絕對的滿意，在出貨後三十天內，都可辦理退貨或換貨。

❹ 若是為您量身訂做的產品，除非有瑕疵，否則我們不接受退貨或替換。

參考答案   MP3 100

❶ Buyer shall promptly notify Seller in writing upon the discovery of any defect. Buyer may return the defective Products to Seller with all related costs prepaid by Buyer first.

❷ Please use these updated questionnaire forms from now on and discard the older versions.

❸ Your complete satisfaction with our products is guaranteed and items may be returned or exchanged within thirty days from when they were shipped.

❹ Where a product has been made to measure for you, unless faulty, we're unable to refund or offer an exchange.

Part
1
國貿基礎篇

Part
2
國貿進階篇

Unit 25

表達弔唁 Expressing Condolence

 國貿經驗談

▶▶ 無論是西方喪禮或東方喪禮，皆屬莊嚴隆重的場合，禮數上採致送輓聯、花圈、花籃或十字花架等，西方習俗上是在卡片上具名贈送者並題字 "With Deepest Sympathy"，而卡片信封上是書寫往生者之姓名，如 "To the Funeral of the Late Mr. Kim Jackson"。就商業關係來說，不太有機會出席國外客戶的喪葬禮，除了距離因素外，主要仍是親疏關係的考量，一般是發函表達對往生者的懷念及弔唁，當然如果商業合作夥伴同是至交好友，則另當別論。

關鍵句型

Sb. be deeply sorry for/that…. 某人對….深感抱歉

例句說明

- We were deeply sorry for the trouble we'd caused.

  我們為所造成的麻煩深感抱歉。

- The company was very sorry for any staff who was effected by the financial crisis.

  公司對於因財務危機而受到影響的

替換句型

- The company felt very apologetic for any staff who was effected by the financial crisis.

express one's feeling to sb. 向某人表達情意

例句說明

- I hardly know how to express my anxiety to the plant about the tight production capacity.

  我真不知道如何向工廠表達我對於產能吃緊的擔憂之心。

- The company should express their regret to any staff who was effected by the financial crisis.

  公司應該向因財務危機而受到影響的每位職工表達歉意。

替換句型

- Our company acknowledges you for bailing us out with economic aid to overcoming the financial crisis.

Dear Olivia,

I were deeply distressed to hear the passing way of your founder, Mr. Kim Jackson.

Mr. Jackson was not only an excellent leader, but also my admired elder. He made great contributions to the success our cooperation during our company operating in the most difficult time. I were especially moved by his solicitude for subordinates and respect to customers. No words can express my sympathy on losing such a nice person.

We all should follow his spirit and keep alive his example.

Yours sincerely,
Wesley Yang

中文翻譯

奧莉維亞 您好：

聽聞貴公司創始人金姆 傑克遜先生與世長辭的消息，深感不安。

傑克遜先生不僅是一位優秀的領導者，而且是我尊敬的長者。在我們公司經營最困難的時期，他為雙方成功合作做出巨大貢獻。令我特別感動的是他體恤下屬，尊重客戶。沒有任何言語可以表達我對失去如此善良者的傷痛。

我們都應遵循他的精神並以他為楷模。

衛斯理揚 敬啟

❶ 電池在 14 天後如有任何異常狀況，雖在產品保固期裡，但仍不符合替換替換品的條件。

❷ 若有每年校正、全面檢查或維修，則此產品保固期可延長 12 個月。

❸ 此保固服務無法移轉，亦不包含定期保養維修、消耗品，以及會正常耗損的零件。

❹ 此保固並不涵蓋因疏忽、不當使用及組裝非原有零件或配件而導致的損壞與故障。

參考答案　　　　　MP3 101

❶ Any battery failure beyond 14 days is not eligible for replacement under product warranty.

❷ Following annual calibration, overhaul or repair, the product warranty will be extended for a further 12 months.

❸ This warranty is not transferable and excludes routine maintenance, consumables, and parts subject to normal wear and tear.

❹ This warranty does not cover damages or defects caused by negligence, improper use, and assembly of parts or accessories that are not original.

職場英語系列 003

國貿英語 E-mail 有一套：我靠抄貼效率翻倍、獎金加倍 (附 MP3)

作　　　者	倍斯特編輯部
發 行 人	周瑞德
執行總監	齊心瑀
行銷經理	楊景輝
企劃編輯	陳韋佑
封面構成	高鍾琪

內頁構成	菩薩蠻數位文化有限公司
印　　　製	大亞彩色印刷製版股份有限公司
初　　　版	2017 年 9 月
定　　　價	新台幣 369 元
出　　　版	倍斯特出版事業有限公司
電　　　話	(02) 2351-2007
傳　　　真	(02) 2351-0887
地　　　址	100 台北市中正區福州街 1 號 10 樓之 2
E - m a i l	best.books.service@gmail.com
網　　　址	www.bestbookstw.com

港澳地區總經銷	泛華發行代理有限公司
地　　　址	香港新界將軍澳工業邨駿昌街 7 號 2 樓
電　　　話	(852) 2798-2323
傳　　　真	(852) 2796-5471

國家圖書館出版品預行編目 (CIP) 資料

國貿英語 E-mail 有一套：我靠抄貼效率翻
倍、獎金加倍 / 倍斯特編輯部著. -- 初版.
-- 臺北市：倍斯特, 2017.09
　面；　　公分. --（職場英語系列；4）
ISBN 978-986-95288-1-8 (平裝附光碟片)

1. 商業書信 2. 商業英文 3. 商業應用文
　　　493.6　　　　　　106013883